© DEMCO, INC.—Archive Safe

COMPLETE PLANS FOR BUILDING

HORSE BARNS

Big & Small

THIRD EDITION

COMPLETE PLANS FOR BUILDING HORSE BARNS BIG AND SMALL

For Information contact:
Breakthrough Publications
326 Main Street
Emmaus, PA 18049

www.booksonhorses.com

ISBN *Complete Building Plans For Horse Barns Big & Small 3rd Ed*   0-914327-91-7

Designed by Kraus & Associates

Edited by Audrey Pavia

Illustrations by Mace Bell, Curtis Gaster, Isabel Guera, Eddie Hartlove, James Hawthorne, Lois Pezzi, Steven Prifti, Kevin Selestok and P.J.Williams.

Printed in Colombia by Cargraphics S.A.

10   09   08   5   4   3

CONTENTS

CHAPTER 5: CONSTRUCTION   105

CHAPTER 6: BARN PLANS   127

CHAPTER 7: REMODELING   233

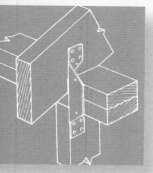

Acknowledgments

Putting this book together has been a delightful and challenging project, much like building your own dream barn. The overall ideas come easy, but the execution is the hard part. So many people have helped, from lending technical assistance, to bright ideas, to plain old moral support, that a complete list is impossible. For one thing, it would have to include a few score of horses who have benefited and/or sometimes suffered patiently as barn inhabitants. Here's a try at listing them:

Carolyn and Patrick Adams, Yarra Yarra Equestrian Ctr., Pleasanton, CA; Hector Alcalde, Takaro, Middleburg, VA; All Afshari, Photo Memos; John Ambrosiano; Amy Breeden Babcox, Warrenton, VA; Hannah Banks, Harrison-Banks Architecture, Boston, MA; Dr. J.M. Bowen, Chesapeake; Virginia Polytechnic Institute and State University; Mrs. John Breeden, Warrenton, VA; Penny and Lloyd Burger; Agnes Clark, Deerpath Farm, Charlottesville, VA; David and Sharon Cooper, Aldie, VA; Luke D. Curtas, Thoroughbred Fence and Supply, Fredericksburg, VA; Linda and Bob Daniel, Brandon, Spring Grove, VA; Kevin Daniels, Manassas, VA; the Dykes, Rockville, VA; Gainesway Farm, Lexington, KY; Dr. J.R. Gardner, Virginia Polytechnic Institute; Steve Gardner, Berthoud, CO; Fred Garrett, Mechanicsville, VA; Gordon Hammond, Bradford, ME; Paul and Phyllis Hassell, Hassell Arabians, Reddick, FL; Mr. and Mrs. Frank Hensley, Breaking Dawn Stables, Keswick, VA; Fritz, Karen and Sacha Himmelmayer, Kafri Farm, Orange, VA; Trip Hoffman and Allen Van Weiren, February Farm, Lovettsville, VA; Dr. Arden Huff, extension horse specialist emeritus, Virginia Extension Service; Carmen Johnson, Windfield Station, Nicasio, CA; Helen and Jack Junkin, Old Mill Run Farm, Mason Neck, VA; George Kindt, Loveland, CO; Fred Kohler, Bittersweet Farm, Middleburg, VA; Klepper, Hahn and Hyatt, Engineers and Landscape Architects, P.C., East Syracuse, NY; Karen Kresge, Connection Newspaper Group, Fairfax, VA; Steve Lamson, Simpson Strong-Tie (R) Co., San Leandro, CA; Doug Linton, Custom Barn Builders, Clifton, VA; Joan Vos MacDonald, Croton-on-Hudson, NY; Helen Makarov, Middleburg, VA; Midwest Plan Service; Gina McRae, Livermore, DA; Doug Meador, Doug Meador Training Center, Lenoire City, TN; Judy Mederos, Mederos Ranch, Livermore, CA; Dr. Bob Mowrey, NC State University Extension Horse Husbandry, Raleigh, NC; Jill Munro, Wedgwood Farm, Livermore, CA; Tolli Nelson, Collie Blue Ranch, North Pole, AK; Jan Neuharth, Paper Chase Farm, Middleburg, VA; Ron and Beth Padley; Christa Panayo, Glendie Farm, Falmouth, VA; Bob Perks, LaGrange Farm, King George, VA; Lois Pezzi, new architectural drawings, New York, NY; Steve Prifti, McGann Saphire, Marion County, Florida extension agent; Alice Quinn, Crazy 8 Ranch, Livermore, CA; Coy and Nancy Sanders, Cutter Ridge Ranch, Livermore, CA; Julie Saxelby, Nicasio Valley Arabians, Nicasio, CA; Joe Ann Scott; September Farms, Ocala, FL; Sherry Shriver; Lalla Rook "Lassie" Tompkins, Briar Patch Farm, Micapony, FL; John Stover, Morton Buildings Inc., Culpeper, VA; Tioga Farm, Parker, VA; Upperville Barns, Division of Northern Counties, Upperville, VA; Dr. H.E. White, Virginia Polytechnic Institute; Doug and Singie Williams, Chestnut Hill Farm, Spring Grove, VA; P.J. Williams Inc., Somerset, VA; Richard O. Zirk, agricultural engineer, Staunton, VA; the horsemen and horsewomen of the Equine-L Internet Discussion Group; and many others.

Forward

Throughout time, horses and humans have enjoyed a special, useful relationship. Today, there is more recreational interest in horses than ever before. Horses are kept for a variety of reasons, including business, sport and pleasure.

In the wild, the horse is an outdoor, forage-eating athlete. In domesticity, people have to provide shelter, food, care and exercise for the horse. This can be as simple or as elaborate as one may choose, but cost effectiveness should be a major consideration. Horse owners invest a lot of time and money in horses, land and buildings. Evaluation and careful planning should be considered in the process of establishing, expanding or renovating facilities.

Establishment of a major facility involves sizable investment and the result, good or bad, is more or less permanent. This book addresses the issues and planning process, and it provides guides to effective horse care and a range of possible facilities. It is an excellent reference for horse owners, prospective horse owners, professional builders, educators and engineers. It would also be a useful text for advanced horse-production classes or schools.

Over the years I have visited hundreds of facilities in all parts of the country. Often, we are asked for advice about plans, locations, renovations and problems. We can never prevent all possible problems from cropping up, but this text should go a long way toward the goal of helping owners and those in the industry.

Early in my career, I was involved in developing one of the first modern-day horse building-and-plans publications. This was later revised, and it is still a useful document, though technology continues to develop. *Horse Barns Big and Small* brings new and well-needed management techniques into focus, and is an excellent guide. Still, the user must visit working facilities, develop a plan, seek advice and draw up a program specific to his or her needs and user intent.

Most owners do an excellent job of housing their horses. However, I generally deal with problems, and there are many, often needlessly repeated. For example, two of the most common issues involve inadequate drainage and poor ventilation. Many stables are dusty and many have inadequate lighting. Labor saving techniques can also be built into the planning process. As for economics, who ever heard of that in a horse operation? But horse farms should be run like any other business. Materials handling, feed, hay, bedding and manure are all too often an afterthought.

Safety for the horse, the worker and the public must be built into the system. Often we are asked after the fact: "How do you like my renovation or location or overall facility?" Frequently, existing facilities should not be renovated or used. Just because a fence is there or a building exists does not mean it is useful. Many farms also have no safe way to control or move animals by planned fencing. Many stables are situated so that if a horse happens to get loose or out of a stall, it has immediate access to a major highway.

Finally, aesthetics, green space, fly control and pasture or paddock management are all important, especially in promoting and marketing the horse.

With all this in mind, we should return to the basics. Why do we have horses? One of the major reasons is to have fun. We often let the process of horse ownership override the purpose of ownership. You need riding and use areas. This book is dedicated to help you enjoy your horses, for fun and profit.

Dr. Arden N. Huff
Professor emeritus, Animal Sciences Department
Virginia Polytechnic Institute
Blacksburg, VA 24061

Introduction

Since you are reading this book, you already own horses or plan on becoming more deeply involved with them. Whatever category you are in, welcome to the horse industry. We use the word "industry" because even if you plan on having horses only for pleasure, you are still part of a vast industry of people who derive their livelihood and recreation from a sport as old as human beings' first domestication of the horse and as young as the delighted child receiving her first pony at Christmas.

Whoever said that the outside of a horse was good for the inside of a man clearly knew the therapeutic value of working with this animal, even discounting time spent riding. A large part of horsemanship is the day-to-day care of the creature—a time-consuming task at best. But while the care and upkeep of a horse requires much daily involvement, the rewards you receive as you survey the world from the top of your favorite mount more than make up for the effort of ownership.

And that brings us to the ultimate purpose of this book. We would like to share with you some ideas that have helped many people enjoy horse ownership. Horse owners all over the country shared their successes and failures, enabling us to put together a book on the best methods of keeping horses.

That's the key to being a good horseperson: seeking out others and learning from their experiences. Not only will you get information to help you with your plans, but you'll likely run into some interesting characters and become quickly absorbed in the local horse community.

In talking to good horsemen and women from Alaska to Florida, we've found they have three things in common. First, they have a genuine concern and like for the horse. Second, they believe that the horse has a job to do and should perform that job well with the correct care and training. Third, they have developed an eye that always looks for the smallest detail, from the amount of water the horse consumes that day, to the quality of its performance, to the overall look of the animal. They incorporate new ideas and experiences into daily encounters with their horses.

Whether you keep horses as a hobby, business or recreation, they can bring you lifelong pleasure. We hope our book will help you become an even more capable horseowner with many, many years of pleasure in the horse industry.

Planning Your Barn

Getting started the right way in the horse world calls for a few key decisions. First, how large will your operation be? Will you have to do the job by yourself, or will you have qualified assistance? And whether you want a small or a large operation, you need to decide whether you want to be in the industry for pleasure or profit.

If you decide to go for profit, your knowledge, management and facility level needs are probably going to be higher, simply because you need to invest money to make money. On the other hand, realize that good management can be used in place of a large capital investment with positive results.

If you are planning a for-profit operation, you'll need a business plan. In fact, even small operations require planning. After all, even if you are in it only for pleasure, you will still reap rewards: the enjoyment you and your family derive from the ownership of horses.

POINTS TO CONSIDER

Whether your plans are large or small, begin planning your facility by asking yourself the following questions:

• Am I Choosing the Right Neighborhood?

While a man's home may be his castle, living next to neighbors who abhor horses can bring on the worst of border wars. If you can choose property in an area where the natives support and share your pastime, do so. The extra amount you might spend for land in a "horsey" area can save you money and stress in the long run. At the very least, select a semi-rural area.

Be sure to check the zoning regulations for the prospective piece of property you are considering. Zoning laws will tell you the type and number of animals you are permitted to keep on a specific piece of property. Check your community covenants, and look for restrictions on barn size and style, the details of manure disposal, and local fencing requirements. Ways to check both zoning and covenants are discussed in Chapter 3.

When purchasing land, look to adjacent areas for riding privileges. Many equine-oriented communities are demanding easements and riding right-of-ways that allow local equestrians plenty of room to ride. Access to trails or land

where the owner has given you permission to ride can prevent you from getting landlocked and riding only in a confined arena.

Don't depend on your realtor to know such information. It's best to get the details straight from the horse's mouth, so to speak, in this case, local government offices. Be sure to check for restrictions on professional riding instruction as well as whether you can legally board horses on the property, and any other factors that may limit your full use of the area for equestrian activities.

• What is the Largest Number of Horses I Can Keep?

The number of horses you can keep may be restricted by the amount of labor available to you, or your own time constraints. Even if you can take care of a large group of horses, zoning laws in your area might forbid it. It is important to ask yourself this question up front if you hope to have a barn that will suit you in the future.

• Will I Do All the Work Myself?

If you plan to move your horses from a full-care boarding stable to your own facility, but have no experience in the day-to-day care of owning a horse, get some first! The effort and work of caring for your horses on a day-to-day basis can be enjoyable, but only if you know what you are in for.

The average working horseowner spends at least 50 percent of "horse time" on barn chores. That's a lot of effort if you're not committed to doing the work. Before you get too far into the project, decide whom the majority of care will fall on—you, a family member, or hired workers. If you are relying on a volunteer, how many other responsibilities does this key worker already have? And how can you make the most efficient barn arrangement so you'll waste the least amount of that person's time?

If you plan to have a child do the barn chores, be aware the amount of work involved can kill the child's interest in horsemanship. You have a couple of options, though, if you'd like to be sure to keep your child's enthusiasm. You can do the work yourself or hire someone to do it. Or, you can design a facility that is as labor-saving as possible and then have a firm family understanding about the responsibilities horse ownership entails.

If this operation will be organized like the traditional rural American family, with all members pitching in, the work load will be easier and more horses can be comfortably kept. However, if you live alone or don't have anyone willing to help you, be prepared to assume all responsibilities for the upkeep of all the horses you plan on owning. This will be a major and time-consuming responsibility depending on how you plan on maintaining the horse or horses on your property.

• What Are My Options in Caring for a Horse?

Depending on your climate and your personal preferences, horses can be kept simply or elaborately. The most simple method of horsekeeping is exclusively on pasture. A good-quality pasture with adequate clean water can supply the total nutrient needs for an average pleasure horse during good weather. If you add a salt block and some protection from the sun or wind, a thrifty pleasure horse can be comfortable with little effort on your part. In this type of environment, your horse is far less likely to develop bad habits from boredom or respiratory trouble from living in an enclosed environment.

During cold or hot, dry weather when grass may be sparse, you may have to supplement pasture grazing with hay and grain. If your local weather varies little in temperature, along with the amount of wind, sun and precipitation you receive, you may be able to manage with trees as a natural shelter. Or, you can opt for a simple run-in shed. Be aware of times when drought restricts water or grass supply and supplement your horses accordingly. Generally, if your fences are safe, your grass and water supply good, and you keep close, regular watch on your animal, pasture is the easiest method of keeping a horse.

If you are actively participating in serious equine competition with your horse, you may feel more comfortable with a stall. Keeping your horse in a stall provides you with more control over the animal's environment. If your horse is stalled, you'll be able to clip his winter coat, reduce summer sun bleaching, and keep a closer eye on possible veterinary problems.

As you can see, you have some choices to make. Should you opt for a simplistic run-in shed, or the more traditional barn/stall arrangement? When trying to make this choice, let yourself dream about what you want, ideally, and then decide what you actually need and can afford. If the property you are buying has existing facilities, take a look at them and see what you have to work with. If you haven't been around enough facilities to have developed a feel for your preferences, go visit some other places. Once you have a sense of what is practical for you, you can get down to the actual project secure in the knowledge that you have done your homework and your facility will be ideal.

Of course, keep in mind that nothing is perfect. A better mousetrap is just around the comer, and what you build today may not always suit all your needs. But if you plan carefully, you'll have a flexible arrangement that can be adapted to most reasonable changes.

Get a firm idea of your needs, develop a plan, and be flexible. Then get started.

COSTS

Figuring costs for your barn isn't hard as long as you deal with ballpark estimates. You can usually build a barn yourself for half the price of contracting it out, provided you count your time and labor as being worth nothing. If your needs are simple and you're handy as a builder, doing the work yourself is clearly the way to go, especially if you take advantage of such prefabricated items as sliding door kits and stock-size trusses.

Most horse owners with more than a run-in shed in mind enlist the services of a barn builder who may or may not have a selection of plans to offer. Most builders will happily tackle any set of plans, of course, provided they are familiar with the requirements and can get the needed materials. If you plan to use your house builder for the barn, be aware that his or her unfamiliarity with barn methods may lead to a higher estimate because a house builder is working from a different perspective than an experienced barn builder.

In general, you can assume that a barn's cost per square foot, as executed by a barn builder, will be about one quarter of the cost per square foot for the average home.

If you go with manufactured packages, you may have less of a variety to choose from to modify your price, but you can count on a generally similar price range. Manufactured barns are very competitive, quite flexible in lay-out due to the modular design, and very popular in many parts of the country. The smaller the barn, the higher the square-foot price, running from around $15 per square foot for a four-stall stand-alone to $12 per square foot for a 10-stall barn from the same company.

Depending on availability in your area, you may be able to save a lot of money simply by good planning. Taking steps like building in 12' increments can allow you to buy all your lumber in that size, which is a standard, stocked dimension and is often sold at a better price per foot than other sizes. Many distributors stock trusses in a 24' width at approximately $40 each, making that an economical choice if you plan to store hay elsewhere. Remember that lumber stores like to deal in volume. If you call for prices on sizes of lumber, be sure to explain the amounts you're dealing with, and see if they will cut you a better deal. If you run short of cash, concentrate on getting the full exterior built first and make do on the inside. Solid doors, hay racks and watering systems can come much later, and your money is better spent on a good price for plenty of pressure-treated lumber.

If you wish to get the most barn for the money, remember that a long barn is cheaper than a wide one. Increasing your roof span will cost more than adding a truss or two on the end. If you want larger stalls than indicated in the plan, you can shrink a center aisle to 10', but no smaller or you won't be able to safely turn a horse around or pass by a vehicle parked indoors.

If building materials in your area are expensive, look into the prefabricated market. Many of the reputable companies that advertise nationwide can work with you to provide a shell for you to fill, or will fill the shell you provide. Even with shipping costs, their bids can be competitive.

Choose materials that are easy to find in your area. For example, a set of plans may call for 4" x 4" posts, but you have similarly sized, pressure-treated poles nearby at a good price. Don't hesitate to substitute materials, provided the new ones are at least as strong, weatherproof and non-toxic as the ones called for.

SOURCES OF HELP

Now that you've made the decision to build your own facility, no doubt you have plenty of questions about your upcoming effort and a vague feeling that there's yet more to think about. You're probably right. Remember that it's wise to seek a qualified professional to assist in the structural design of any barn you are planning.

If this is your first time building your own facility, there is only one way to learn—by asking questions. Find help with your project, but remember there are three kinds of help: Free help that is worth something, free help that is best ignored, and valuable help you pay for.

After you get a little bit of experience, you'll be able to identify free help that turns out to be useless. The horse industry is full of "experts" who are only too glad to tell you everything they know. Keep in mind that taking bad advice may get you into trouble. Conversely, an enormous number of knowledgeable horse people are out there and willing to offer good advice and assistance. Time and experience will help you learn the difference between the two.

Many equine magazines offer excellent articles on horse management and facilities. Subscribing to one or two of the better ones will give you some fine ideas for the day-to-day working of your barn and management of your horses.

There are many professional horse people who have moved into the world of consulting and can help you with everything from selection of the right property or horse to international marketing, and everything in between. Farriers, veterinarians and feed dealers can be great sources of help. Another excellent source of good, free information are the local county government offices serving the agricultural community.

One of these, the Cooperative Extension Service (CES), has a local office in nearly every county in the nation.

Originally started in the early part of this century as an extension of the land-grant universities, CES's purpose was to improve the quality of life for those living in rural environments. Today, its purpose is to disseminate information free of charge (or at nominal fees) on a wide variety of agriculturally related subjects, many of which involve keeping and maintaining horses. This information is available to the public.

CES staff members, called county agents, extension agents, or agricultural, and youth and/or home economic agents (depending on your state) can give you first-hand information on pastures, facilities, feeds, feeding, fencing and a host of other topics. They also have access to the knowledge bank at their state agricultural college. Many extension offices have expanded their services to include information on farm planning, record-keeping and computer programs for farms. All in all, the Cooperative Extension office may have so much information to share with you that it may take you several visits to receive the full benefit of its services.

A federal office whose job is related to land management and whose office is often located near or in the same area as the Cooperative Extension Service is the Natural Resources Conservation Service (NRCS). Originally called the Soil Conservation Service, it is a service of the U.S. Department of Agriculture and has offices in every state, with staff members who have extensive knowledge about soil and soil usage in their jurisdictions.

Knowing if the land you are purchasing (or already own) is suitable for livestock and pasture, and the best location for your house, your roads and your barns, is valuable information. The NRCS office can offer you assistance in identifying the type of soil in your area, its best uses according to its profile, and how you can overcome some of the not-so-desirable parcels you may be forced to deal with. Consulting NRCS can save you money from the start. Forewarned is fore armed.

Finally, another government agency that you may or may not need, depending on the scope of your operation and whether it is for pleasure or profit, is the Consolidated Farm Service Agency (CFSA). Originally called the Agricultural Stabilization and Conservation Service, the CFSA provides a streamlining of services provided to those in agricultural endeavors. It may well be worth your time to find out about the CFSA services in your area if you are serious about the horse business. The agency's function is to regulate the dissemination of government funds for the agricultural industry in this country. Funding exists for assistance in establishing and maintaining certain crops and pasture lands in given regions, and your farm may be eligible if it meets the CFSA criteria.

Don't overlook your county or city planning department and other local government offices that are concerned with land use. They often have staff members familiar with pertinent subjects. Local community colleges, if they have agricultural programs, are also a fine source of help, as is the Federal Department of Parks and Recreation in your area and the National Forestry Service.

A word about government bureaucracy: The Cooperative Extension Service and the Natural Resources Conservation Service offices have few or no strings attached to their services. The Consolidated Farm Service Agency offices are more tightly regulated and require more paper work, as money is often involved in their transactions. You'll have to decide if the financial rewards are worth the effort in dealing with this office. Be sure to ask questions. As an ordinary horse owner, you may not qualify for certain awards, but there are some funds available on a limited basis that can be of significant help to your operation.

Organization is key to sorting through all the information you will get. By organizing the data you've gathered as to type and source, you'll eventually build a file that can help you put together a coherent picture.

Set up your files according to the categories of information you'll need to build your facility. Some suggestions include but are not limited to:

 Fencing

 Barn plans

 Construction materials

 Ventilation

 Plumbing and water sources

 Electrical wiring/lighting

 Arrangement of inner-barn facilities (feed rooms, wash pits, tack rooms and so on)

☑ Farm layout plans

☑ Government (federal, state, local) offices offering help and contact people

☑ Same list for regulations

☑ Reliable sources of information such as knowledgeable horse people, governmental agencies and consultants

WORKING WITH BUILDERS

If you see the construction of a full-sized barn as too time consuming a task, plenty of professionals and talented amateurs can take over for you. However, before you turn your land and your checkbook over to them, be sure you see eye to eye on what's being built.

A word of caution if you're also building a house: The people who build your new house, unless they are also farm folks, may not necessarily be your first choice as barn builders, merely because you are asking them to build a structure that's outside their area of expertise. Basic as it is, a barn has requirements of strength, drainage and ventilation that a builder inexperienced in agriculture will have to think twice about.

If you have a barn builder in your area, look into this person's rates. Consider hiring the builder as either the person in charge on the site or as a consultant. When functioning as a consultant, the builder can ensure that boards are always nailed to the horse's side of the post, that swinging doors open in the correct direction, and that all the other details peculiar to the barn are done right.

The barn builder will generally give you the best price because he or she knows the local costs. A homebuilder, used to estimating such things as drywall and concrete slabs, may be dealing with too complex an equation for a "simple" barn. A barn builder may also help you save money as he or she will be aware of the types of materials and structural designs suitable for your area and your needs.

Once you've selected your plan and determined your general material and labor costs, you are ready to proceed yourself or approach a builder seriously. Before you start building or before you select a builder, remember to set up your files and think about your facility. Spend time determining what you want in a barn before you make any phone calls. Decide which extra frills you want, the essentials you need, and any future changes you might consider. No builder will expect you to know everything about the barn you want, but you do need to give him or her something solid to work with.

If you discuss your dream barn with a builder and find the cost is beyond your means (keeping in mind that getting a building project done "under budget" is well-nigh impossible), ask if the builder can be hired on a "time and materials" basis. This means the builder can work to a certain level of finish based on what you can afford. This is not as convenient as having a turnkey job, in part because the builder may not be available right away when you are ready to resume the job, but it allows you to build in stages and work toward your dream barn. Here again, planning the job in stages, either with the builder or alone, can save you money in the long run.

When evaluating professional builders, get at least three estimates. The estimates will give you a ballpark figure, and you'll probably find that the prices differ widely. Before going for the lowest bidder, look over what the builder is proposing. Examine the material types and levels of quality being offered, as well as the construction methods each uses. This way, you are comparing apples to apples in your builder selection.

At some point, you'll need to get serious about the firm, bottom-line budget for your project. Be realistic, and know that your estimated costs and the actual ones may be pretty far apart. Be prepared to compromise if you don't have a big financial cushion to fall back on. Building a smaller barn, or the first stages of an expandable one, may help. Develop the shell you want, but hold off on the water heater, tack room carpeting, automated bug sprayer and the like. Whatever your choice, preplanning will help you feel more satisfied with your barn at each stage in its development.

If you aren't up for the job yourself and there are no barn builders available in your area, take a homebuilder on a tour of some well-done barns. Impress upon him or her that the structure must be horse-proof; a horse's needs must come first in a barn. Overestimating is as useless as underestimating, of course, so be prepared to give your builder a realistic picture of life around a barn.

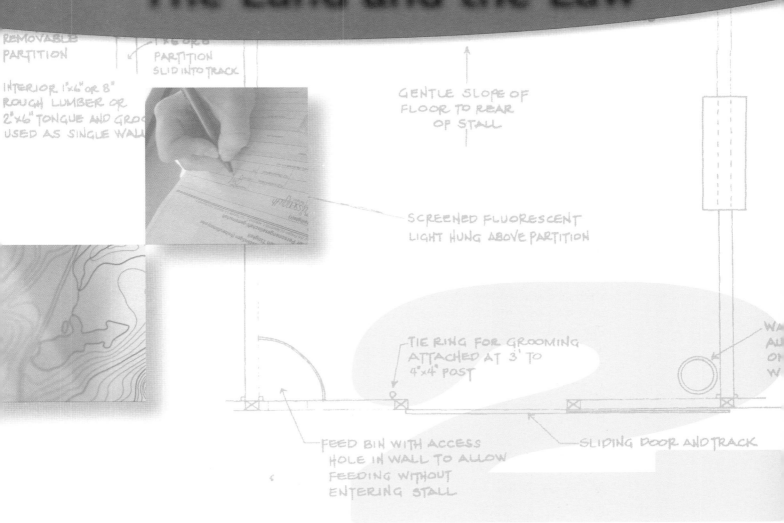

REMOVABLE PARTITION

1"x6 OR 8 PARTITION SLID INTO TRACK

INTERIOR 1"x6" OR 8" ROUGH LUMBER OR 2"x6" TONGUE AND GROO USED AS SINGLE WALL

GENTLE SLOPE OF FLOOR TO REAR OF STALL

SCREENED FLUORESCENT LIGHT HUNG ABOVE PARTITION

TIE RING FOR GROOMING ATTACHED AT 3' TO 4"x4" POST

WA AU OF W

FEED BIN WITH ACCESS HOLE IN WALL TO ALLOW FEEDING WITHOUT ENTERING STALL

SLIDING DOOR AND TRACK

Before you buy a single board for your barn, prepare to conform to a myriad of local requirements, even though what you're building is "only a barn." If you think zoning requirements are only for houses and factories, you'll need to bone up on agricultural and recreational land use in your area. Zoning and building codes are governed by the state, and also by the local county or city in which you build your barn.

ZONING ISSUES

Some communities may not have any requirements beyond those established by the state for land use and building safety, which are part of an effort to ensure quality of life of their communities. Other regions have strict requirements to the basic zoning and building codes. The key is to know what your area allows or demands, specifically with regard to keeping livestock and the buildings that go along with this lifestyle. Before you start, learn about the kinds of permits that are called for before and during construction. Plumbing, electrical work, footings, and drainage all come under an inspector's jurisdiction.

Most states and counties divide up land according its preferred use. General classifications are:

 Agricultural

 Residential

 Commercial, or business

 Industrial

These categories can be further divided so the land you own may be tightly restricted as to the number and kinds of animals allowed, what they're used for, where on the property their housing is placed, how it's fenced, and much more. The local zoning office can tell you exactly what you can and cannot do with your land.

Never take anyone's word about the zoning on a piece of land, and don't make assumptions about the zoning by looking at what is around you. Always go to the main source: the zoning commission.

In addition to regional limitations on horse keeping, your community may have its own covenants—restrictions designed to maintain a consistent appearance in particular neighborhoods. Also, the person or developer selling the property may place some restrictions in the deed. Read this document carefully and consult an attorney if you are in doubt about any sections. By protecting the neighborhood, you are also protecting yourself from others who might want to move in and run an undesirable operation close to you.

Covenants, deed restrictions and even zoning can sometimes be waived if the reasons are compelling enough, but the bureaucratic tangle can be daunting. Ask yourself if the land is truly worth the trouble before you start, as waiver negotiations can slow you down considerably.

As we all become more aware of the impact of our actions on the environment, it's important to plan your facility to be in sync with the best interests of the land and nearby waterways. Whereas this once was a matter of courtesy, it is more and more becoming a legal matter: Are you near a waterway, a drainage ditch, a high water table? The runoff, or leachate, from your manure is full of nitrogen compounds and suspended solids that can have a major effect on the streams and underground water supplies of your area. Horses are now recognized as producers of "non-point source pollutants," making them subject to control by the Clean Water Act and under the possible jurisdiction of your State Water Resources Control Board and other agencies. The Environmental Protection Agency may even get involved.

Before you do a lot of planning, check with the Natural Resources Conservation Service, the Regional Water Control Board and any other local agencies that supervise land and water use in your area to see if there are steps you should take to protect the environment around your barn. These agencies may recommend using buffers or filter strips of plantings between your manure areas and waterways, static composting in controlled bins, or even weekly pickups from a dumpster.

Consider developing a land management plan, possibly with the help of your local Cooperative Extension Service or the Natural Resources Conservation Service. This will make it easier for you to meet federal, state and local conservation measures, as well as making your property more efficient and productive. Ask if you are eligible for any federal, state or local funds to help put your plans in action. While a grant may not fully fund pasture reseeding to stop erosion, up to 75 percent of the cost may be available to help with the project. It never hurts to ask, and you'll be doing everyone downstream a big favor.

Once you are aware of the restrictions on your property, you can plan a compliant farm design, and then file for building permits. Here again, the range in regulations extends from "no permit required for agricultural buildings" in areas zoned for them, to barns needing the same code inspections as those required on a standard suburban home.

Call or visit your local building inspector's office to see about "blanket" permits that give you permission to go on with more than one small project at a time. While the aggravation of dealing with all these regulations for a simple, uninhabited structure may seem too much, if you work with, instead of against, the inspectors and planners, your entire project will be simpler. In fact, some barn builders, in anticipation of problems with inspectors who want house-type construction used in every case, have had good luck when they've made the inspector almost an advisor on the whole project. This way, they benefit from the inspector's local experience on land, materials, workers, etc., and they avoid an adversarial relationship that can hold up the extras, such as lounges and bathrooms.

PLANNING ACREAGE

You will want to have pastures for your animals, if at all possible. Horses, by nature, are grazers, and do better if they have the opportunity to move and eat constantly as is their natural inclination. This provides them with daily exercise while reducing your feed costs.

In areas with dense populations, high prices, and lack of available ground for pastures, horses are maintained very successfully on small plots. This type of horse keeping requires good horse-management practices, as close confinement can increase problems such as internal parasite infestation or sand colic from eating off bare ground.

Erosion and water runoff are problems with small dirt paddocks unless their location is chosen carefully. Use a small strip of grass as a buffer around the paddock to help prevent rapid water runoff. Consider applying mulch to high-traffic areas, such as around fence lines, to reduce dust.

Since the conditions for pastures differ not only from state to state but from county to county, it's important that you check with local experts to get the best information on your soil and climate type. Help can be found from the Cooperative Extension Service or from private farm consultants who can assist you with mapping a pasture plan specifically geared to your situation.

Animals Per Acre

The number of animals recommended per acre of land is a variable that can fall into such a wide range that it's best to consult local zoning laws. In areas where the climate is warm year-round and the rainfall ample, you may actually be able to figure one horse per acre. The other extreme is a climate that is so dry and sparse that it takes five or more acres per animal to provide any kind of nutrients.

Generally, if you live in a seasonal climate area and plan on feeding hay and grain in addition to pasture, estimate two acres per horse. During warm months, you may be able to cut supplemental feeding to nothing, depending on the work your horse is doing and how easy the horse is to keep.

Variety of Grass

Through experimentation, the agricultural industry has developed a large variety of grasses that can be targeted to withstand drought, high rainfall, dense soils, close grazing and cool as well as warm seasons. Getting expert help on what to plant in your pasture can enable you to plan the right variety for your specific situation.

For horses, not just any old grass will do. Grass requires judicious care and constant upgrading. Some grasses are not appropriate for horses. For example, many common types of grass known as fescues carry a fungus that causes abortion in broodmares. If you buy land for a breeding operation and have acres of this grass, you will need to completely reseed if you want to eliminate worry. Other grass types, while harmless to horses, are not strong enough to take the pressure of constant grazing. They can go from green to extinction in one season.

If you are establishing a pasture for the first time, you'll find it difficult to wait for your pasture to mature after seeding and seeing the green grass grow. General recommendations are to seed in the spring or fall and to keep livestock off the pasture until it is truly well established—not just looking a little green. This may take a full year or even two, depending on the climate and type of grass you plant.

Failure to wait until your grass has put down a strong root system will result in your horses pulling the new tender grass shoots up, roots and all, thereby leaving nothing for new growth. Good pastures, having an established root system with a reserve of nutrients to allow refoliation, enable the leaves to replace themselves as they are grazed. As a rule, horses should be removed from a pasture when plant height is reduced to 1- or 2".

Stretching Your Pasture

One method of maximizing the use of your pasture is to plant both warm and cool season perennial or annual grasses. If your varieties are compatible, you can use the property year-round.

The best method for stretching pastures is management. The following management techniques will add life and quality to your pasture.

ROTATE ANIMALS. If at all possible, rotate your horses on the available pasture land. Sectioning off the pasture is a cheap way to force your animals to completely graze an area. If you cannot afford the additional cost of fencing, use inexpensive electric wire.

Horses are selective grazers and will chew down certain areas, creating "roughs" and "lawns." These lawns then become the newest growth places and contain the tastiest and most tender grasses. They will generally defecate in one area of the pasture, the rough, and then not eat from that area, which is actually a form of self-protection from over-infestation by worms.

Knowing this can also help you plan your healthcare program. Patrol your pasture area weekly, if not daily, and pick out the manure piles just as you would in a stall. It's a lot of work, but it will pay off in better health for your horses. You can have the horses fully graze a portion of pasture and then move them on to the next small area while the first recovers properly. The result will be fewer parasites, more evenly grazed fields and better overall management.

FERTILIZE AND RESEED. After removing the livestock, put your management skills to work as you lime, fertilize and mow the pasture for maximum regrowth. The best time to fertilize is in the fall, not early spring, as fall feeding directs the new energy into the roots for stronger plants. For spring feeding, go with a higher phosphorus/potassium, lower-nitrogen fertilizer to encourage further root development instead of a sudden rush of green that stresses the plant. Although an extra-green spring pasture looks nice, it will not be healthy when drought and extensive grazing tax its endurance.

All soils eventually become depleted of the nutrients necessary for plant growth unless you resupply them. After all, the soil's nutrients go into the grass and then into your animals. To see exactly what your pasture needs, take a soil sample, have the extension office send it to a land-grant university, and get an exact reading of your soil. In the absence of a soil test, apply 60 pounds each of nitrogen, K20 and P205 plus 2 tons of lime per acre. Some companies provide custom application of fertilizers and weed killers. If you have one in your area, call for a recommendation.

If you have an established pasture that is not being used for hay and has high levels of K20 and P205, re-fertilize with those nutrients every three to four years to maintain their levels in the soil. If clover represents at least 30 percent of the plant population in the pasture, no nitrogen fertilizer is needed. But if clover is absent, apply 50 to 60 pounds of nitrogen in early spring and again in late summer each year.

If your established stand of grass thins out, reseeding is called for. Do this in February or March, using a no-till drill to seed into the existing thin sod. Do not allow horses to graze these pastures until the old plants reach a height of 5- to 7". When the old plants are grazed down to 2", remove the animals again to let the new growth establish itself.

CONTROL WEEDS. Another important aspect of pasture management is controlling and eliminating weeds. You can do this by either mowing or applying chemicals, with a combination of the two being the most effective practice. Mowing must be done before weeds go to seed.

Chemical applications should be timed to eliminate weeds as they emerge with full leaves in the spring, but not so late that they are strong and well-established. When shopping for chemicals to control weeds, always read the labels. You can poison your horses if you don't watch out. If you need in-depth information on a particular chemical, check with the local Cooperative Extension Service for local recommendations. Commercial chemical and fertilizer companies can also provide information and custom applications of their products.

In most cases the chemicals you purchase for your farm will be readily available to you over the counter. However, for some highly toxic chemicals or those you need to purchase in large quantities, you may find that you need a pesticide applicator's license.

You can readily obtain one by going to your local Cooperative Extension Service office. You will receive information on specially scheduled classes and on taking the pesticide applicator's license test (usually open-book). Upon completion of the test and its scoring, you'll receive a rating enabling you to purchase certain types of pesticides not available to untrained buyers.

Whether or not you plan to use large quantities of these types of chemicals, the test and classes are useful as they teach the correct techniques for storing, using and disposing of chemicals on your property or at approved dumping sites. With the renewed awareness of hazardous materials and their disposal problems for the environment, it is wise to update yourself on regulations for your area and to know the hazards of storage and use that could threaten your health or that of your family or livestock.

A word of caution about chemicals: While there are many products approved by the Food and Drug Administration (FDA) with directions that sanction immediate grazing, it is always best to allow rain or heavy dew to wash the chemicals into the soil and off the grass. Most tests for toxicity apply to oral ingestion, but do not address possible damage to a horse's lungs from inhaling chemicals.

REMOVE MANURE. Removing manure is your best method of parasite control as mentioned above, but it is also the one you are least likely to keep up with over the long haul. Horses will pick an area of their pasture in which to defecate and then won't graze that area. Consequently, pasture grasses will become uneven. This condition will only get worse if nothing is done to correct the situation.

Traditional recommendations used to be that you clip and drag the whole area to level the grasses and spread the piles. Current research indicates that the life span of internal parasite eggs is such that it allows them to lie dormant, so they can re-infest your horse, even if you allow the pasture to weather for a month or more. Some researchers feel that dragging only spreads the eggs in a wider area, thereby increasing re-infestation. Realistically, most horse operations do not have the option of leaving pastures unused for long periods of time.

While the most effective solution is work-intensive, it is the best for your horse's health. You should treat your pasture like a stall and pick the piles out of it as often as possible. Experiments with heavy-duty lawn vacuums and lawn brooms have not yet produced the perfect result, but a gentle stroll around the pasture with shovel and wheelbarrow can do wonders. Also, an effective oral worming program for your horse is a must.

PREPARING YOUR SITE

No matter the shape of your land, plan to put your building on higher ground, even if you have to build it up with fill or clear some trees to get to a high spot. The goal is to create a slope away from the barn or run-in shed that will carry waste and rainwater away from the footing. If you don't do this, waste and water will rot out the wood as well as make your barn a sloppy mess.

A run-in shed should drain just as efficiently as a fancy barn. Both of these simple and complex facilities have two common functions as far your horse is concerned: to protect the horse from wet weather and to encourage the run-off of urine. Placing the facility in the highest location is especially important if your soil drains poorly.

Be sure to use treated lumber for your barn footings to avoid rot. Even with treated wood, the daily wear of horses in and around the stable can take a toll on the structure and the surrounding area. If your finances allow, it's a good idea to lay a rough-surfaced asphalt or concrete apron around the entire barn at least 4' from the base of the building and make it slope gently away from the barn. Too steep a slope can invite accidents, so allow just enough of an angle to draw water downward, about 1" per foot.

A simple way to create a concrete apron effect is to slope a layer of stone dust out from the barn, then sprinkle powdered cement dust over the stone. Wet it liberally from the hose, or wait for a rain, and you'll find the stone dust dries as a solid.

If this is more than you want to undertake, run downspout drainage well away from the barn. Even barns need gutters. If pastured horses have access to the barn, you will need to horse-proof the drainage, as horses like to loaf around the corners of a barn and can crush lightweight drainage pipes that lie close to the surface. As with any part of your barn, make sure your plans do not compromise your horse's safety.

One solution to the drainage problem is to buy clay drainage pipe, 4- or 6" in diameter, and place a section at the bottom of the downspout, which runs into the end of a small trench leading away from the barn. Then run black corrugated plastic drain pipe to the end of the trench. If you are on a hill and the horses have access to the end of the pipe, make sure you use clay pipe where the pipe resurfaces. Otherwise, the end will be flattened by the horses and the pipe will back up to the base of the barn.

If your barn is midway down a hill, level the ground above the structure or even put a shallow ditch between the drainage and your building. Then steer the runoff well away from the barn so it won't be undercut in the first big rain.

If you're building in marshy territory, you'll have to dig a series of ditches and dry wells to overcome a damp barn area. You can have these dug by an earthmoving company at some expense, or you can easily rent a small ditch-digging machine that will excavate a ditch with a minimum investment of time and money. You can then place black perforated drain pipe in the ditch, which can be covered over, thereby eliminating an open ditch in the field. If you calculate the slope of the ditch properly, the underground drainpipe will pull excess water from the area.

To put a dry well in a troublesome low spot, dig a hole at least the size and circumference of a 55-gallon drum and fill it with brick and stone debris. Dig a ditch nearby, run a small slanted trench from the dry well to the ditch, and place a length of corrugated black plastic tubing from the top of the well to the bottom of the ditch. Use the type with scattered holes throughout. This will speed drainage in all directions.

Finally, plan your buildings with an eye to esthetics. Keep the front of the property pretty, landscaping trampled areas behind the barn, and your farm will look more professional.

Full Farm Planning

I takes more than a lone barn to make a farm, and you can ease the work required to run your facility by good placement of the important extras: paddocks, rings, roads and storage buildings.

Each item you wish to build on your property will have unique requirements as to footing, drainage and access. So before you build, decide what extras you are going to have, and then where you will place them.

For the horse show on the go. A little structure like this is very handy for putting on horse shows, whether on your own property or someone else's. This compact announcers stand was designed with the racks in the rafter for a sound system that can be operated off a car battery or with an extension cord to the nearest power source. It can be unhooked, leveled and set up in minutes. With a few cinder blocks inside for a ballast, the trailer can then be hauled down the highway to the next show.

As shown, the trailer's a bit small for a show secretary or a dressage judge's use, but the idea is excellent and could be scaled up to provide room for two, given a trailer base of a slightly larger size. More and more facilities with room for competitions are using portable judging stands, secretary booths and even cross-country jumps, built on heavy runners that can be dragged to new locations with a tractor—it's a good way to keep your property fresh and interesting as a show site, and you have unlimited flexibility to grow through the years without wasting materials. Photo by M. F. Harcourt

GETTING STARTED

Begin with a diagram of your property—either the actual site plan or something that includes the major features of the terrain—and add the fixtures to scale. This will give you a fairly accurate idea of the relationships between these features and their proportions to each other and your overall property.

Indicate on your map the type of soils located on the property. Your Natural Resources Conservation Service representative can help you map your property as to soil type and use, and may even be able to make recommendations as to the best building sites and road and pasture locations. Since soils have different profiles and thus different best uses, you may realize that your first choice for a road site may be a wet, soft piece of soil at the driest of times, far better suited for pasture than for a road. And if you plan a bathroom in your barn, your soil must have even better percolation than if you planned only a drain field.

Your goal is to get the most for your money. If you have a natural area for a roadbed, why pay money to place the road in a location that may require tons of gravel? Only you know your overall plan, and it may well be that the lay of the land and other considerations force you to place the road, ring or barn in a certain location. Even facilities of less than two acres can contain plenty of amenities if they are well planned.

This is where overall planning pays off. Being aware of extenuating factors beforehand will keep later problems from stopping your whole operation in its tracks.

Once you've located the best spots for the barn and any outbuildings, you can plan the driveways and fencing. Here

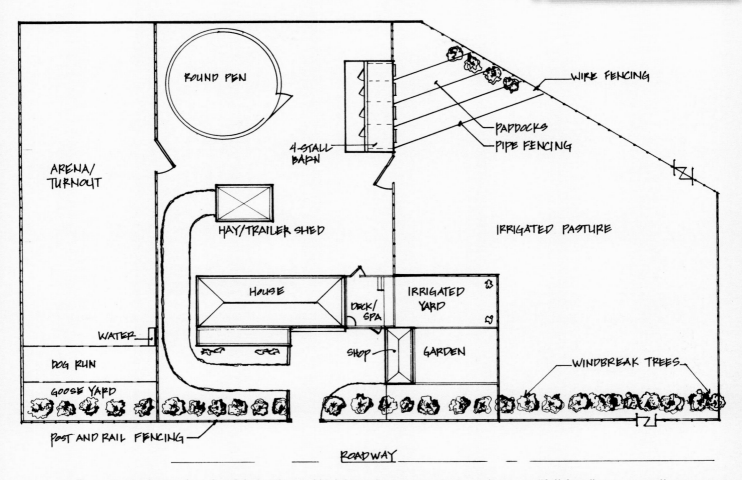

Two-acre ranchette plan. Careful planning and high intensity management can give you a "full farm" on very small acreage.

your key thoughts should be access and security: access to every part of the farm for vehicles large and small, and security for your horses from each other and passersby. Unless you are working on the tightest of budgets, or with a highly unusual site, plan from the start to enclose the barn area with its own fencing, separating it from regular pastureland and public roadways.

Fencing the barn away from pastured horses will save you from such problems as chewed siding, rubbed corners, crushed drainpipes and much more. Unless you plan a true loafing shed, don't let your barn be treated like one. Loafing horses often turn into pests.

They cannot resist harassing other horses who, once stabled, cannot escape the annoyance of a wandering companion. Horses loose in the barnyard can also find their way into tack and feed rooms, damage parked tractors and trailers, and steal loose grooming implements. Worse, they can escape from the barn area and onto roads and others' property. While fencing your barn will mean more expense and putting in at least one more gate than you would need for pastures alone, the security is worth it.

Before planning the gate sites on your property, draw up your preferred driveway plan. Don't think about casual visitors when setting up the driveway,

Using a water tub to serve adjoining fields saves refill and cleaning time. Here, the owner has also thought about keeping birds from perching over the tank and soiling the water. A strip of short roofing tacks offers little comfort for resting foulers of the water. (Glendie Farm, Glendie, Virginia). Photo by M. F Harcourt

Old bathtubs are tempting as field waterers, but they have some drawbacks. Placed as shown, the sharp edges can damage a horse's knees with just one stamp at a fly. One option would be to frame the tub in with a skirt of 2"x4" lumber covered with plywood. Photo by N. W. Ambrosiano

but of delivery trucks. Trucks need the greatest turning room and most secure footing. Since your deliveries will be to hay and grain storage area, bedding areas and perhaps manure pickup spots, the driveway should be in as straight a line as possible from the street to the end of the barn. Add a large enough pad or backing lane so that even an 18-wheeler can make a three-point turn. Consider building a circular drive around the site. As soon as you order a few tons of hay, you'll understand the advantage.

When planning your driveway, don't skimp on the width. Trucks and trailers track wider than a car. If the driveway's

shoulder is steep, you risk a serious accident even at slow speeds if a tire should drop over the edge.

If your barn is on a slope, take the expense of grading your road into account when you plan both your barn and the outbuildings. A few dozen more feet of road can cost a great deal. Perhaps moving the barn a touch closer could save dollars better spent on a new saddle.

Managing the Neighbors

In today's world, horses are often looked upon in the same legal sense as swimming pools and other elements that appeal to young people: they are considered attractive nuisances. The very presence of a horse is like a magnet to many children, and you should keep this in mind when selecting and planning your facility. Setting your fence lines several feet inside your property's

borders allows you to ride around the perimeter and also buffers you somewhat from the neighbors.

Since any normal fence can be easily scaled by a child, consider placing "No Trespassing" signs in strategic locations on the property's borders. Develop a rapport with neighbors who have children so they understand they are more than welcome to visit when you are home, but that your property and animals are off-limits at other times.

Become a good neighbor by realizing that not everyone enjoys the smell of horses or the sight of a neat manure pile. Considering the following suggestions:

☑ Keep the overall appearance of your property up to neighborhood standards. Keep the pastures clipped and clean and the barn painted or stained.

☑ Keep trashy-looking areas to a minimum, including piles of used lumber or jumps. If you must store building supplies, do so neatly and in a less-visible place.

☑ Isolate manure to a confined area and remove it regularly to keep down insect populations.

☑ Implement an ongoing fly and insect control program so that neither you nor your neighbors are bothered by the pests.

☑ Keep dust levels to an acceptable minimum by following good pasture management and applying an economical surface treatment to your road and work areas. Again, both you and your neighbors will benefit.

☑ Take time to drag rings and riding trails to keep unsightly manure piles to a minimum.

In small areas such as paddocks that need a larger water supply than a simple bucket, plastic garbage cans are very useful, especially if secured to the fence by a rope or chain through one handle to prevent tipping. They can be easily cleaned, are cheap to replace, and will not hurt a playful horse. (Wedgwood Farm, Livermore, California). Photo by N. W. Ambrosiano

☑ Reduce a potential eyesore by efficiently storing equipment and machinery in your facilities.

☑ Plan solid footing for areas of high traffic so these areas will withstand vehicular and animal movement without creating mudholes.

☑ Arrange large service deliveries at times when they won't disturb neighbors by blocking the road and making excessive noise.

☑ Limit your farm pets, and allow for space to confine your dogs. Since many dogs like to harrass horses, the safest approach is to keep your dogs restrained.

☑ Consider landscaping that is not only aesthetically pleasing but also easy to maintain. If you can, keep your yard small, and turn all available land into pasture.

The way you lay out your farm should make your operation more efficient and easier to work around, as well as more pleasant for you and the neighbors.

MANURE DISPOSAL

Most people have no idea how much manure a big, healthy horse can produce until it snows so deep you can't get the wheel barrow out of the barn to dump what you have cleaned from your horse's stall. One of the foremost problems for horse keepers is manure storage and its ultimate disposal.

If you or your neighbors like organic gardening, you have the biggest headache essentially solved. Horse manure makes fine garden mulch once it is well aged, and with proper liming, it can enrich the poorest of soils. However, it does need to age several weeks before planting because of the high ammonia content.

Storing manure until it is disposed of is another story. Criteria for adequate manure storage should include the following:

☑ Manure storage should be close enough to the barn to be convenient, yet far enough from the barn to avoid trouble with flies, and far enough from the house not to create odors in hot weather.

☑ The manure storage area should be easily accessible from the barn, no matter the weather. If your property is on a hill, you'll find it's easier to push a full wheelbarrow downhill and an empty one uphill.

☑ Manure should be contained in some manner so that it does not become like the eggplant that ate Chicago and engulf the stable yard.

☑ The containment and eventual disposal of the manure must comply with local, state, and federal (Clean Water Act) regulations regarding watershed and ground water contamination. Manure is on the list of "non-point source pollutants" and falls under the control of such agencies as the State Water Resources Control Board and the Environmental Protection Agency.

Spreading fresh manure back on pastureland, even if you plan on leaving the land vacant for several months, is not a good idea, as the eggs of various species of internal parasites have great longevity and can re-infest your horse months after being spread. New studies have proven that the old theory of spreading manure so as to expose eggs and larvae to sunlight and air simply does not kill adequate numbers of parasites. Thus, you are only re-infesting your horses with these methods.

You have other options. First, manure can be composted in an enclosed structure (built with cinder blocks or other material not eaten away by the manure) and sold or given away in the spring to avid gardeners.

The traditional composting method has been to spread a coffee can of slaked lime over each wheel-barrow load of manure you dump, keeping the pile stacked up well and adding as much as possible to the top of the pile, rather than the front.

A newer composting method is becoming popular amongst those who wish to sell high-quality compost with quick turnaround time. Called Static Bin Composting, it involves a series of bins with fronts that come off and several lengths of perforated PVC-pipe that can be laid across at intervals.

As each bin is started, the pipes laid across allow air in to speed the composting of the inner layers. With a series of bins cooking away, you can open the front of the most-cooked bin when the compost is ready to sell without disturbing the batches that still need time to complete the process.

The aerobically composted material has less odor than other types, the bins keep your stable area tidier, and should inspectors from local agencies come calling, it's clear that you are doing your part to responsibly treat the manure produced.

No matter what type of composting you choose, be sure that the approach to the pile is not fouled with spilled bedding. This will create an unsightly and unsafe path in a heavily traveled area of the barn.

(Note: Research at North Carolina State University has provided interesting insight into the role of lime and the microorganisms essential to good manure pile composting. To provide the swiftest composting environment, hold off on applying the lime to the fresh manure until you are ready to spread the aged pile on the fields, as the organisms prefer the high-acid environment of the manure. Once they have done their jobs and reduced your pile to a well-rotted compost heap, then apply the lime to more closely balance the acid level before you use it as fertilizer.)

You can spread dirty bedding on pastures not intended for horses, if you are not endangering any waterways near the field. Most parasites are species specific, so cattle can benefit from the enriched grass you provide without picking up the parasites your horses would get.

You can also spread manure on ground intended for crops, but, here again, check with your Cooperative Extension Service as some crops do not tolerate this well. In some areas, manure containers and removal companies will handle it like any other garbage—a fine arrangement, although potentially expensive.

BEDDING STORAGE

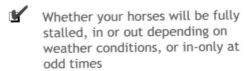

hoosing bedding of the type and amount you require depends on several factors:

- ☑ Whether your horses will be fully stalled, in or out depending on weather conditions, or in-only at odd times

- ☑ Whether you'll be cleaning all the stalls or have reliable help or hired labor to do it for you

- ☑ Whether bedding is easily available, economical, and aesthetically and otherwise satisfying

Once again, your source of information should be the Cooperative Exten-

The owners of this barn used the same brick from their house to build their barn so the two buildings would match. Photo by M.F. Harcourt. (Private farm, Southern Pines, North Carolina)

sion Service and other horse owners in your area, as they will know the types of bedding most available in and best-suited for your area.

In addition to straw and wood chips, consider chopped newsprint sold in bags, or sand, peanut hulls, peat moss and pine needles. Each has its own pros and cons, but all share one characteristic: They must be kept dry to be effective. As a time saver, if bedding materials are stored along the path to the manure pile, you can pick them up in a wheelbarrow on the way back to each stall.

Bedding materials that are stored in piles, such as wood shavings, require frequent truck deliveries, so you need to provide room for a dump truck's raised bed, which is ordinarily blocked by most shed roofs. To solve the problem, consider a roof that can be lifted off in sections, or one that can be raised from one side with a pulley. That way you'll avoid dragging snow-laden tarps back and forth, or worse, filling a stall with wet bedding.

INDOOR ARENAS

No matter the climate, an indoor arena is a great boon to a barn owner who has either a boarding or training operation. Shelter from sun, rain and snow allows consistent exercising and a secluded atmosphere for working young or excitable horses. Building your own arena is a task that won't be detailed in this book, as the clear span of an indoor arena—at least 66' x 132' for the smallest

We recommend that the bedding be stored in a separate shed from the barn, both for dust control and fire safety reasons. Using a separate storage area, you are then free to develop a design that fits your budget and your stable area style—from two-story, pole buildings with cupolas to a plywood bin with a lift-off roof panel, you can arrange the most reasonable plan for your operation.

Another innovative idea is to have a rolling roof on the storage bin, one that's suspended above a roller track. At Deerpath Farm, they built a stall-sized shed 12' x 12' and 8' high. Behind the shavings shed is an equipment shed of similar size but 1' shorter in height, which serves as the support table for the sliding roof and extended track when drawn back for dump-truck deliveries. Photos by M. F. Harcourt.

The roof has a chain extending from the front center to allow a person to grab it and walk straight back toward the back of the shed, pulling the roof back at the same time. This worked well for the men designing it, but the barn owner, a woman of small stature, found it difficult. Instead she prefers to use less upper-body strength and places a 2" x 4" board, 8' long, against the third rafter and walks forward, pushing the roof into retracted position instead of pulling.

The builder, Hugh Markin of Albemarle Barns in Charlottesville, Virginia, notes that the sliding roof idea is limited by the size of shavings area to be enclosed and the amount of snow the roof must support. If too large or heavy a roof is planned, it will either bind the rollers or take several people to push out of the way, defeating the purpose. In order to accommodate a slightly heavier arrangement, one could use heavy-duty rollers that hang vertically from their track and suspend the roof under them. Photos by M. F. Harcourt.

Shavings on the farm built by Penny and Lloyd Burger of Chesapeake, Virginia, are kept dry yet accessible with this roomy container. Made of 4" x 4" posts with plywood sheets as walls, it is roofed with a series of lift-off panels that allow stand-up or drive-in access as far back as you wish. There's no need to drag a heavy tarpaulin or move an entire cover all at once. Roofing panels can be fiberglass or corrugated aluminum over 2" x 2" framing, strengthened as needed as the dimensions of your shavings pit require. Photo by M. F. Harcourt

This two-sided shavings pit features a simply framed metal cover that is hinged on one side. Pulleys are attached to the free side and to the barn near the ridge pole, allowing one person to easily lift the shavings-pit roof for deliveries. The two-sided design with raised cover, unlike a three-sided one, lets a dump truck drive through without backing to unload. (Windchase, Hillsborough, Virginia). Photo by M. F. Harcourt

of dressage arenas—is beyond the skills of most amateur builders. You can have either an arena company or any builder with the expertise for such large buildings do everything from just the roof and its supports to the plushest of turn-key arrangements, including a wet bar in the lounge if you want it.

If you decide to add an indoor arena, remember that having a steel-building contractor raise your roof has several benefits. First, you'll get a column-free interior, which is an absolute prerequisite for a safe indoor riding area. Second, the roof does not require the support of the walls, so you can either go without walls, or make them as weather tight as you wish. Just remember that you are building a year-round facility, and if your weather ranges from subzero to 90 degrees F, you must allow for both insulation and good air flow.

If you wish to finish the building yourself once a contractor has done much of the work, your options are almost unlimited. In warmer climates, you may choose to merely fence the covered area. This gives you a working area that is sheltered from the sun and rain and still allows summer breezes to blow through it.

If inclement weather is a problem, then walls are called for, but it is possible for them to be weather-tight without creating a claustrophobic feeling. When using any exterior ply product available or noisier sheet metal over a simple 2" x 4" frame, be sure to add several 5'x 8'windows on each side and at either end. You can put a sliding glass door over each, or a shutter arrangement, but either way, you'll have the shelter of the indoors with the light and air of the outside. A panel of light, fiberglass-reinforced plastic or Plexiglas at the top of the walls will give you a much lighter indoor arena as well, without compromising your weatherproofing.

If you plan to insulate the walls and ceiling, note that swallows, pigeons and sparrows are among the many noisy squatters who will move in if you don't prevent it. Stretch wire mesh across every bit of insulation or you'll have a bird condo from wall to wall with so much noise that you won't be able to hear a trainer from 10' away. Rigid, foam polystyrene (Styrofoam) insulation or sandwiched, insulated roof panels are good bird-proof alternatives.

No Footing, No Horse

The footing indoors is as subject to wear and tear as that in your outdoor arenas, even if it doesn't get direct rain and wind. Many types of special arena footings are advertised in horse magazines, or you may have your own favorite. Most of them combine wood shav-

An entire shed for shavings is not out the question if your operation is large enough. This shed combines closed storage, at right, with two-bay shaving and/or vehicle storage under a 14-foot roof. (Paper Chase Farm, Middleburg, Virginia). Photo by M. F. Harcourt

ings with another material, such as sand or peat, for springiness, freeze-proofing and reduced dust. Companies that supply footing materials can dump the basic material for you to install, or you can have them prepare the footing from below ground level on up.

The essential ingredient in all these footings is a good base that drains well and neither sinks deep nor packs down to a stone-like surface. There are as many combinations of sand and gravel suggested by experts as there are experts, so preparing your footing gives you a prime chance to talk with local excavators, agricultural advisors and other arena owners about what works best in your region.

No matter what kind of top dressing you put over the base of your arena, it, like your horses, will take some grooming to be at its best. Lack of rain produces a dusty arena that leads to coughing, so you'll need some irrigation plans. If you don't mind spending several hours with a hose in your hand, you can try hand-watering. A step above that is installing lawn sprinklers upside down along rafters, which provide a more automatic arrangement (for example, you can set these on timers to sprinkle an hour before the first ride of the day). This also gives you some wet patches, though, which are unacceptable in a potentially perfect riding surface.

The smoothest water devices have copper tubing with tiny holes every foot or so that are laid out in a grid or rows along the rafters the length of the arena. They are operated, like any sprinkler system, from a main valve and work best if you have good water pressure to force an even mist over the full area.

Dampness alone won't give you great footing. It can be overdone, and in cold weather, it can make even the most freeze-proof surface pretty icy. Under such conditions, limit your watering to moisten the top few inches of the surface; whatever you do, don't drench it heavily with water.

Raking or dragging is the other big job required to maintain arenas, especially those with heavy use. Hand-raking all the way round the edge is effective, but it uses time you'd probably rather spend riding, pulling manes or even cleaning tack. If you have a farm tractor, or even a small car with a trailer hitch, you can help maintain any footing by running a harrow or shallow cultivator over it every week. This breaks up clods, erases trenches and keeps the footing aerated well below the surface, which is essential to protect your horse's legs.

Circular, rotating harrows with short teeth are wonderful for arenas, but they require a three-point hitch on a tractor, as do most liftable harrows. Simple drags, while not as sophisticated, can do a decent job and don't require the tractor hitch. A chain harrow, for example, has a mesh of heavy links with protruding fingers on one side and a smooth surface on the other. Held square by a

heavy pole at the front, it has a chain that hooks on to any pulling vehicle so it can be dragged rough- or smooth-side down, depending on the final polish you want on the arena's surface.

In a pinch, a sheet of wire-mesh or chain-link fencing will work, provided it's weighted down with a railroad tie or two and held square at the front. This will not provide deep conditioning, but only smooths out the rough bits.

Extras

Once basic walls, footing and conditioning are planned, you can consider the nonessentials. For a start, a sloping interior wall, 4' high, will keep your horse from fracturing your kneecaps as it attempts to evade your inside leg in the course of your schooling sessions. A gentle slope of about 12 degrees moving outward from bottom to top will ensure that as the horse's feet come close to the bottom wall, you still have a foot or so of space before your stirrup begins dragging along the higher wall. This interior wall must be made of heavy boards or thick plywood, built over reinforcing 2" x 4" planks. Otherwise, a misbehaving horse or one turned loose can kick the wall while passing, and might even force a leg through it with disastrous results.

Along other safety lines, don't allow any objects to protrude into the arena, such as jump poles stored alongside, door latches that stick out, or hooks for hanging training equipment. The walls at horse and rider height should be completely smooth and safe. Don't forget to provide a tractor-width door to the inside, not just a horse-width one.

Lights are essential for any arena with walls, even for open-sided arenas where people like to ride after dark. You'll have to invest in some powerful hardware; the lights must hang fairly high in the air, not require constant maintenance, and spread light evenly across the floor. A walled-in arena gets very dark very quickly, even if you install clear plastic panels along the eaves. Metal halide lighting provides a brighter, whiter light, which is even suitable for videotaping under, but costs more initially. Don't hang lights lower than 15' from the floor, and if you plan any jumping indoors, take them up to 20' or more to be safe.

Check your local suppliers for their best buys on lights big enough to illuminate a wide area; you don't want spotlights. Remember that you're lighting a dark-walled, non-reflective room, so increase your estimated wattage requirements to account for that. Placing sodium-vapor, street-type lights every 20' or so is effective, but check with local builders for their recommendations.

It's good to have a viewing gallery, no matter what your planned use for the arena. This area can be as simple as an 8'x 16' judges' stand that raises occupants 3' off the ground, or as elaborate as a heated seating area complete with popcorn and plush seats. From a viewing area, buyers can get a better look at "the product." If the viewing area is closed off but has a window, discussions and comments won't intrude on the rider's concentration. If you include a window, make sure you can open it to call instructions to your riders.

BREEDING SHEDS

A special breeding area is essential to some farms, whether a fully booked season calls for all-weather servicing or mere modesty requires an enclosed area. Like an indoor riding arena, a breeding shed needs good, dust-free footing, plenty of headroom and good lighting. More than that, though, it needs a wash stall and teasing area to prepare both mare and stallion. The wash stall can also double as a lab and examination station for the vet.

Solid construction is essential for any teasing barriers and breeding chutes as well as plenty of padding on exposed corners and edges. Prefabricated metal chutes may be ideal for your situation, or perhaps you want to design your own.

A word or two about the layout of barn-to-breeding-shed is handy here. If you are trying to show a breeding stallion, keeping his two jobs clearly separate can be a challenge. If you can arrange your facility so that the breeding shed is in the opposite direction from the riding and grooming area, the stallion is likely to be far better behaved as he leaves his stall. He will know from the outset his job for the day. Having a special breeding halter and shank that are markedly different from his regular tack also helps clue him. It's easier to insist on proper manners when the horse's hopes haven't been raised unnecessarily.

The separation of riding and breeding areas carries through to wash stalls and grooming sites as well. Install such areas in your breeding shed and don't use them for anything else so the stallion is never led astray by your choice of a grooming site.

If you're thinking of standing a stallion from a breed that permits artificial insemination, you're going to need a "phantom" mare for collecting semen, and a laboratory to process it. We won't go into the details of setting up a top-quality lab. A vet who specializes in breeding operations can you give you ideas on this, as can a vet school with breeding classes or a visit to a facility you'd like to emulate. That said, here are some ideas for a startup arrangement.

One new breeder in Pleasanton, California, developed the phantom mare shown on page 45 to withstand the affections of a 17-hand warmblood. It took the help of a professional welder and the use of an old telephone pole to get a safe, solid start.

Have a welder attach a cradlelike bracket for holding the telephone pole to each of two 6-foot-long-heavy iron pipes, at least 5" in diameter. Sink the pipes approximately 4' into the ground, surrounded by a thick ring of cement. Your final depth will vary according to the height of your stallion, but in this case, the top of the telephone pole, once several inches of padding is attached, is 54" off the ground.

Lay the telephone pole into the cradle brackets and bolt it into place firmly. For additional stability, you can cross brace the pipes to one another with chain, but if you choose to do so, pad them so the stallion's leg cannot be hurt

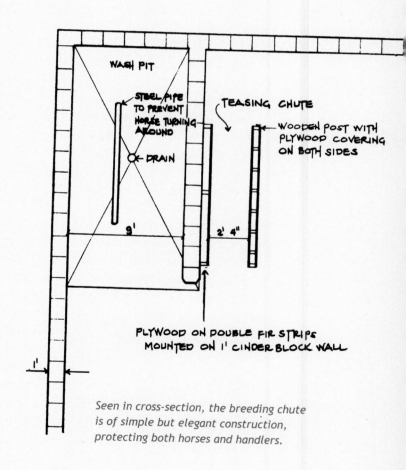

WASH PIT

STEEL PIPE TO PREVENT HORSE TURNING AROUND

DRAIN

TEASING CHUTE

WOODEN POST WITH PLYWOOD COVERING ON BOTH SIDES

9'

2' 4"

PLYWOOD ON DOUBLE FIR STRIPS MOUNTED ON 1' CINDER BLOCK WALL

1'

Seen in cross-section, the breeding chute is of simple but elegant construction, protecting both horses and handlers.

if he misplaces it.

Add heavy padding all around the pole, and fill out the "waist" of the phantom to an approximate 24" diameter. The final layer of material needs to be something similar to canvas that will not tear if it is bitten or rubbed by the heel of a shoe, but it shouldn't chafe the stallion's knees or belly either.

A cutout wooden mare's head and neck (mane and eyelashes optional) can be attached to encourage your stal-

lion's affections, and be sure to place the phantom's ears facing forward so she presents an inviting silhouette. For one stallion, "Farah Fantom" has a brown canvas body, palomino neck and head, and white mop mane.

In the case of the breeder in Pleasanton, the laboratory became the site for further innovation, as the temperature requirements of the artificial vagina (AV) and the sperm are quite precise. To heat water for the AV above the 120-de-

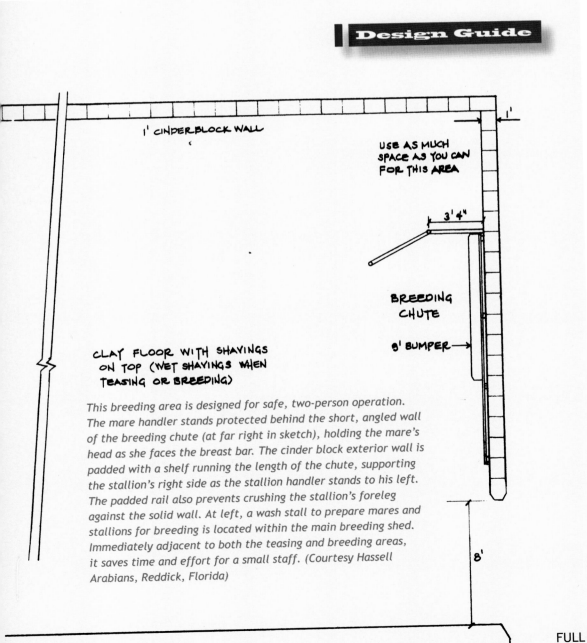

Design Guide

1' CINDER BLOCK WALL

USE AS MUCH SPACE AS YOU CAN FOR THIS AREA

1'

3' 4"

BREEDING CHUTE

8' BUMPER →

CLAY FLOOR WITH SHAVINGS ON TOP (WET SHAVINGS WHEN TEASING OR BREEDING)

8'

This breeding area is designed for safe, two-person operation. The mare handler stands protected behind the short, angled wall of the breeding chute (at far right in sketch), holding the mare's head as she faces the breast bar. The cinder block exterior wall is padded with a shelf running the length of the chute, supporting the stallion's right side as the stallion handler stands to his left. The padded rail also prevents crushing the stallion's foreleg against the solid wall. At left, a wash stall to prepare mares and stallions for breeding is located within the main breeding shed. Immediately adjacent to both the teasing and breeding areas, it saves time and effort for a small staff. (Courtesy Hassell Arabians, Reddick, Florida)

Breeding Farm Walk-thru Safety Checklist

Answer each question with a yes or no, then evaluate the responses and make improvements where needed.

STABLES

A. Electrical System

1. ❑ Is each building's service entrance equipment located in dry, dust-free location?

2. ❑ Is service entrance equipment mounted on fire resistive material?

3. ❑ Is service entrance equipment free of rust and other signs of deterioration?

4. ❑ Are electrical fixtures properly covered so they do not fill with cobwebs, dirt chaff?

5. ❑ Are circuits properly fused with correct size fuses?

6. ❑ Is all wiring in good condition with no signs of fraying or deterioration?

7. ❑ Are all lighting fixtures properly protected?

B. Heating Units

1. ❑ Are heating units designed for use in barns and stables?

2. ❑ Are heating units properly insulated from combustible material?

3. ❑ Are heating units serviced at least annually by a competent heating contractor?
(Date last cleaned and serviced _____)

4. ❑ Is the use of portable heating units prohibited?

C. Above-ground Fuel Storage

1. ❑ Are tanks located at least 40 feet from any building?

2. ❑ Are tanks properly grounded?

3. ❑ Is 5 pounds or larger class ABC fire extinguisher conveniently located near tank?

4. ❑ Is tank properly protected from collision by vehicular traffic?

D. General

1. • Are areas surrounding buildings free of high weeds, grass and debris?

2. • Is hay properly dried and cured prior to inside storage?

3. • Are all roofs, walls, windows and doors weather-tight on hay storage buildings?

4. • Are fire extinguishers:
• Located in each building
• At least 5 pounds ABC or better?
• Conspicuously hung within 50 feet of any point in the building?
• Protected against freezing?
• Inspected and tagged annually?

5. ❑ Are lightning rods properly installed and grounded with conductor cable showing no signs of corrosion?

6. ❑ is there a responding fire department within 5 miles of the farm?

7. ❑ Is there an evacuation plan? Are all staff familiar with it?

8. ❑ Is telephone number of fire department conveniently located near telephone?

9. ❑ Are NO SMOKING signs posted and enforced?

10. ❑ Are horses valued over $100,000 stabled in several buildings to avoid catastrophic fire loss in one building?

MANAGEMENT

A. Pastures and Paddock Areas

1. Describe size of pastures and paddocks, and type of fencing:

❑ Pastures: _____
_____
_____

❑ Paddocks: _____
_____
_____
_____

2. ❑ If post and rail type fencing, are rails secured to inside of post?

3. ❑ Are pastures/paddocks free of harmful objects?

4. ❑ Are isolated groups of trees fenced off or protected by lightning rods?

5. ❑ Are pastures rotated to break the life cycle of parasites?

6. ❑ Are shelters provided in pastures/paddocks?

7. ❑ If yes, are roofs free span?

B. Stable Areas

1. ❑ Are stable aisles well lit and at least 12 feet wide?

2. ❑ Are stable aisles and walls free of objects that might harm horses?

3. ❑ Are all stalls designed to prevent contact with neighboring horses?

4. ❑ Are all stall doors equipped with horse-proof latches to prevent escape?

5. ❑ Are all electrical fixtures/wiring inaccessible to horses or properly protected?

6. ❑ Are stalls cleaned and rebedded daily?

7. ❑ Is all grain and feed kept in covered containers/bins?

Credit: American Bankers Insurance Group, Larry A. Lawrence, Animal Science, VA tech., Blacksburg, VA.

gree limit of the barn water heater, the breeder brought in a restaurant-style coffee urn. With the addition of a few cups of cold water to the preheating supply, the urn's water is about 130 degrees, which cools slightly in the AV on the way to the breeding shed, and is just right at the preferred 123 degrees when the stallion is ready to produce.

To keep the AV, collection bottles, gel and everything else that touches the sperm warm enough, a light bulb was installed on a large insulated Coleman cooler. The bulb is shielded to provide heat but no light, with a rheostat for temperature fine tuning. Thermometers are in place in the cooler and water heater to keep close track of temperature, as this is critical to every phase of collection and storage.

Your lab will need more than a coffeepot and a cooler, and that's when veterinary supply catalogs and an eye

"Farah Fantom," the phantom breeding stand shown, is an example of the solid construction it takes to develop a workable artificial insemination station. Designed with steel posts set in concrete, cross braced with chains and a telephone pole set in an iron cradle, it withstands the affections of a 17-hand warmblood stallion. The head and mop-maned neck serve more than a purely decorative function; they, along with the general dimensions of the phantom, serve to keep the stallion's mind on the project.

on the second-hand market can come in handy. A good second-hand microscope for checking sperm motility, for example, can save you hundreds of dollars.

TRAILER & GENERAL STORAGE

The variety of vehicles and support equipment that goes along with horse keeping is surprising. Storage of this equipment can be simple, and if properly organized, can add years to the life of your machinery. By properly storing your trucks, trailers, tractors and other equipment, you not only protect your investment but ensure safety by keeping them away from your animals.

All vehicles that use combustible fuels should be stored in facilities away from your barn. Storing vehicles adjacent to the barn increases the danger of fire, as well as the potential for respiratory damage to your horses from fuel exhaust. Even using tractors for manure removal while the horses are in the barn is ill advised. If this is absolutely necessary, open all doors and windows to allow maximum air circulation while work is underway.

In building an equipment shed, you'll find the sizes that work for horses are almost perfect for vehicle widths, but not for lengths. An average tractor is 6- to 8' wide, with a length of 12- to 16'. A manure spreader can be as narrow as 5', but rarely wider than 7', although its length can be as much as 16'. A truck needs 8' of width and 24' of length, and a trailer needs at least that. Be sure to add extra space to open vehicle doors, load and unload saddlery and conduct minor maintenance.

All equipment should be stored so that the horses cannot come in contact with parts that may cause injury. This goes for lawn mowers, small wagons or any other heavy equipment. Implements used in your daily cleaning operation in the barn itself should be stored up out of the traffic patterns of both horses and humans for additional safety. Be sure to store your wheelbarrow in an out-of-the-way location, too.

A trailer shed to hold these items— just a simple pole building with optional sides—can also double as a hay storage shed. Hay can be kept in the loft or on the floor area. The hay's dust particles will do no harm here, and you'll get two types of storage under one roof.

FENCING

The possibilities for fencing are almost endless these days, since the technology of plastic, vinyl-and-wire, rubber and metal fencing has been well developed. And of course, there's always wood.

Good fences truly make good neighbors, or they at least keep you from having one more reason to disagree with the neighbors: loose horses. Your enclosure must be secure, economical, durable and safe for errant livestock. At one time, only post-and-rail fence or a combination of post and wire mesh fit the bill. While those are still a highly popular methods of fencing, new materials are gaining popularity because of their durability, ease of installation and safety.

Depending on your situation, you can pick the type of fencing best suited to your budget: slip rail or board, post and rail, wire mesh withaa board or electric wire along the top, rubber, smooth high-tensile wire, wire with vinyl "rails"

This is a good example of a workable trailer, hay and run-in shed. (Captain and Mrs. D. A. Williams, Chesapeake, Virginia.) Photo by N. W. Ambrosiano

attached, metal poles, PVC poles or rails, or plain electric wire. They all work.

The only wire you should never have anywhere near your horses is barbed wire. Horses can kill themselves on such fencing. Even the quietest old campaigner's instinct is to run until stopped, and then struggle when tangled. Combine those instincts with a fence that breaks the skin at the slightest touch, and you have mangled horses.

Board and woven-wire fences have posts spaced at 8-foot intervals, while high-tensile wire posts are located every 30 feet, with droppers at 10-foot intervals. Material costs vary with terrain, source of material, and other factors.

You can add to the lifetime of your fence by following some simple steps during installation. First, set every post as deeply and firmly as you can. This means tamping the soil to a concrete-hard texture around every single post. If your soil is damp and rots wood quickly, soak the bottoms of all your posts in creosote for at least a week before setting them in the ground, and then surround them with a sand mixture to let the area drain well. (Creosote usage is regulated by the federal government and many states, so to use this chemical you'll need to contact your local environmental protection agency.) Even better, substitute PVC or metal posts for wood in marshy areas and take advantage of their impervious properties.

If you're using a fence that takes tension, such as high-tensile or rubber, follow the manufacturer's directions to the letter; your fence will stay up and function far longer than if you hadn't. You may need an extra tool or two, such as

Split-rail fencing is handsome and secure in areas where it is available. Left unpainted or stained, it eases labor. Like post-and-board fencing, it is excellent for areas of rolling terrain or uneven pasture borders. Photo by M. F. Harcourt

For regular use, a 3-board post-and-rail fence, nailed with the boards on the horse's side, is excellent. Like 4-board fencing (pictured) this can be used for winding fence lines and on hilly terrain as each panel stands independently. (Takaro Farm, Middleburg, Virginia) Photo by M Harcourt.

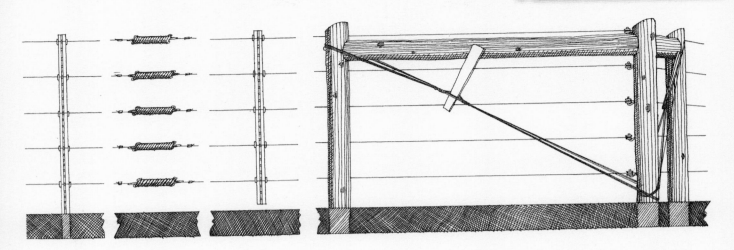

A high-tensile wire fence is excellent for large properties. The long distance between support posts makes it economical, and, once the corner braces are properly set, the fence lasts a long time. Any or all of the wires can be electrified to keep horses off and intruders out. (Techfence, Advanced Farms Systems, Bradford, Maine)

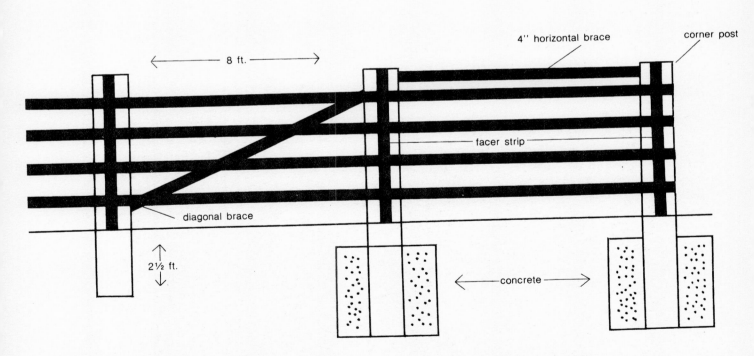

Rubber fencing has received little press since its initial splash into the market during the 1970s, but it makes a wonderfully economical, long-lasting fence. At installation, some important steps are required in addition to firmly placed posts. A propane torch is needed to seal off edges where loose strings may attract chewing horses, and a facer strip of fencing run down each post helps hold things together longer. (CFM Inc., Marysville, Ohio)

A spring-loaded roller is one method of keeping vinyl fencing tight. (LaGrange Farm, Bob Perks, King George, Virginia). Photo by M. F. Harcourt

Vinyl fencing on wooden post can be both attractive and functional. (LaGrange Farm, Bob Perks, King George, Virginia). Photo by M. F. Harcourt

While PVC fencing is expensive initially, its long life and easy care plus elegant beauty may make it a bargain in the long run. (LaGrange Farm, Bob Perks, King George, Virginia). Photo by M. F. Harcourt

This 4-board rough-cut oak fencing is well matched by a safe, solid pipe gate. The hinge post is a 6" x 6" that can take the stress of a gate's movement over the years, while the latch post need only be a 4" x 4" matching the rest of the fencing. The gate is centered between the posts with plenty of clearance, but not so much that pushy adult horses can trap themselves. (Takaro Farm, Middleburg, Virginia) Photo by M. F. Harcourt.

In its open position, this gate appears a little unusual. But for the sloping pastures of this northern California ranch, it's a custom-made item that's just right. Rather than having to find or grade out a flat area for each hillside's gateway, a local pipe-fence company welded a series of angled gates that match the slopes exactly. (Cutter Ridge Ranch, Livermore, California) Photo by N. W. Ambrosiano.

A cheater gate, or dropout rail, is a good option for an area where you need occasional access but prefer not to install a gate. Rails slide through brackets on each post, and a peg fits into a hole through both bracket and rail to ensure that horses don't rub the fence open. (Brandon, Spring Grove, Virginia) Photo by M. F. Harcourt.

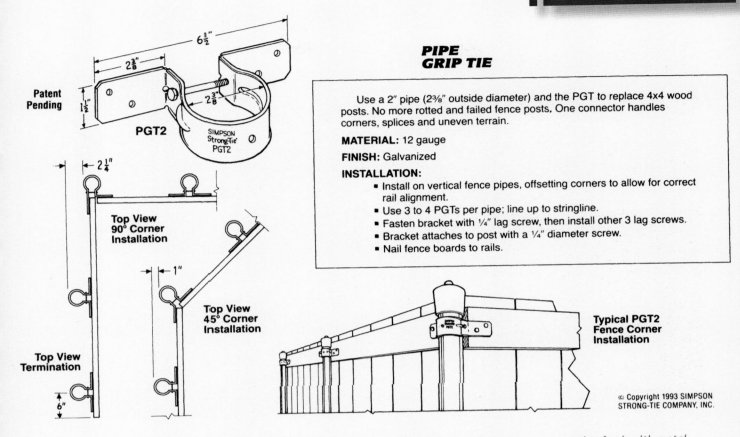

PIPE GRIP TIE

Patent Pending

PGT2

SIMPSON Strong-Tie® PGT2

Use a 2″ pipe (2⅜″ outside diameter) and the PGT to replace 4x4 wood posts. No more rotted and failed fence posts. One connector handles corners, splices and uneven terrain.

MATERIAL: 12 gauge

FINISH: Galvanized

INSTALLATION:
- Install on vertical fence pipes, offsetting corners to allow for correct rail alignment.
- Use 3 to 4 PGTs per pipe; line up to stringline.
- Fasten bracket with ¼″ lag screw, then install other 3 lag screws.
- Bracket attaches to post with a ¼″ diameter screw.
- Nail fence boards to rails.

Top View 90° Corner Installation

Top View 45° Corner Installation

Top View Termination

Typical PGT2 Fence Corner Installation

© Copyright 1993 SIMPSON STRONG-TIE COMPANY, INC.

Not all wooden fences have to have wooden posts. If you are replacing chain-link fence or starting fresh with metal posts in mind, specialized hardware exists to attach boards to the posts. If you plan to use chain link, keep in mind that the top edge of the fence needs to be shielded from contact with the horses either by a metal pipe or wooden rail attached as shown. (Courtesy Simpson Strong-Tie Company, Inc.)

Tired of her young warmblood stock grabbing at each other over the fences, yet reluctant to double-fence the pastures, dressage trainer Jill Munro had these 7-foot post-and-board fences built with diamond-mesh wire stretched to cover all the wood surfaces. There are no sharp edges to hurt rubbing horses, it's too hard to play over easily, and the wood fencing is an effective barrier for the horses that they can't mistakenly run through. For security purposes, this fencing is excellent. It presents a challenge for potential horse thieves who might otherwise snip a wire or pop a board out of place to steal stock. (Wedgwood Farm, Livermore, California) Photo by N. W. Ambrosiano.

Gates can either be centered between the posts to swing two directions, or they can overlap to meet the post securely from just one direction. This is especially handy for paddocks where small foals might be turned out, as they are geniuses at entangling themselves in small spaces. (Wedgwood Farm, Livermore, California) Photo by N. W. Ambrosiano.

Regardless of fence types, horses standing by paddock and field gates tend to wear through the footing. Using sand over a packed gravel bed, or even a rubber stall mat as shown, can make the gateway less of a mud hole when the rains come. (Mederos Ranch, Livermore, California) Photo by N. W. Ambrosiano.

For the safest pastures, round off your corners so there's nowhere a horse can trap another for a kicking contest. (Albemarle County, Virginia) Photo by M. F. Harcourt.

Pass-throughs like this one at Glendie Farm in Virginia, currently owned by Bill, Ling and Kaja Baum, save wear and tear on gates and the people opening them. Designed with a reinforcing bar at the top to keep pushy types from forcing their way through, a pass-through should be just wide enough for a person (including winter coat) to slip through sideways. These are not advisable for farms with small ponies or foals, for obvious reasons. Photo by M. F. Harcourt.

Even beside buildings, a small section suitable for people and not horses allows you to come and go without the aggravation of opening and closing gates. (Brandon, Spring Grove, Virginia) Photo to M. F Harcourt.

Where pasture land is at a premium, such as in many parts of California, small paddocks adjoin the stalls. While it's more pleasant for the resident horses, they do get bored and territorial, fighting over the dividers and kicking through the bars. To solve this without stringing a web of electric fencing, Jill Munro and veterinarian Andrew Clark designed a series of extra-large dividers that project above and beyond the regular fencing. Here Grand Prix dressage veteran Sol, at 17 hands, looks over the back fence of his paddock. Visible behind him are the 2" x 8" kick boards bolted between the partition railings up to a 4-foot level, and you can see that the divider extends far enough to prevent neighbors from reaching around the end of the paddock to harass each other. (Wedgwood Farm, Livermore, California) Photo by N. W. Ambrosiano.

Gates that can be opened from horseback are a great advantage, provided they are hung solidly and aren't allowed to sag out of position. This system allows a rider to pull back the top handle, which is held in place by a metal track along the top of the gate, and the bolt slides through a guiding sleeve into the gate post. This is a simple system to build from metal or wood gates. To prevent horses from accidentally triggering this switch, an L-shaped safety catch is just visible at the edge of the handle. It is actually a bolt bent to a 90-degree angle with the nut welded to the gate top, rotating out of the way with the flip of a finger. (Cutter Ridge Ranch, Livermore, California) Photo by N. W. Ambrosiano.

A wire fence with an electric top wire is often used when horses and other livestock are kept together. A solar charger means any pasture can use an electric fence for added safety. (Ken & Helen Montgomery, Ashley Heights, North Carolina) Photo by M. F. Harcourt.

Stone walls, in regions where stones abound, are handsome but not the safest for horses. They shift over time, becoming easy to climb over, and a careless horse playing in a field can be injured by crashing into such a wall. If your farm has stone walls, simply line them with an electric wire to keep horses clear, or run a separate line of fencing inside the perimeter. Photo by M. F. Harcourt.

A Note on Wood Preservatives

Creosote has long been used to preserve fences and to keep horses from chewing on it. While there were some discussions in previous years about pulling it off the market for environmental reasons, the issues seem to have been resolved. Creosote can be found in a number of fence painting products and is used by many fence companies as the cover of choice for its black color, resistance to decay and lack of palatability to horses.

Two newer preservatives are CCA and ACQ. Both products give preserved boards and post a green tinge to the wood grain. Again, safety discussions regarding the arsenic used to preserve the CCA lumber have caused companies to create a product with far less arsenic and much more copper, making the ACQ lumber products a better choice for you and your horse's safety.

As of November 2004, costs of a four-board CCA fence with half-round post cost $2.93 a linear foot, according to General Timber, Inc. of Sanford, North Carolina, a major fence building company. The same fence in the ACQ preservative runs $2.59 a linear foot. Both prices include installation.

Remember that price quotes fluctuate with the market and locality you're in. When inquiring about cost, ask if installation is included in the quoted price. Buying directly from the mill may save you money but the lumber may still need to be treated or painted to preserve it. Putting in fence posts and getting the boards on correctly is very hard work. This may be an area in your farm plans worth the extra money it would take for the professionals do it.

posts, or sometimes every post, in concrete. That is not difficult, but it does add to overall fencing costs. Other incidental costs to consider are shipping fees from fence manufacturers, specialized fasteners for synthetic fencing, and paints, stains or coatings for wooden fences.

To keep board fencing at its best, consider the following: If fencing with oak, don't paint it. Oak lasts twice as long if left to weather gently on its own. Other woods benefit from a coat of paint or stain applied when the fence is dry. To keep rot away, leave the bottom edge of each board bare. This allows trapped moisture to escape.

Apply paint with a brush or sprayer, coating each board thoroughly. Missed spots let in rain, and the board will rot inside the paint coat. When using creosote, you can get the fastest coverage with the type of heavy-duty, 4" wide roller used in applying textured paints. These hold up best on the rough fence surface. Be sure to wear rubber gloves, goggles and a face mask when applying any of these chemicals, as none of them is good for you. Some will react with the sun on your skin as you work, giving you a nasty rash, too.

a stretcher or special tension-holders, but these items are worth the expense. A fence that is set correctly the first time is always stronger than one you have to fix along the way.

When you install your fence, place the posts on the outside of the fencing materials so horses cannot run into the hard edges of the post.

With all but post-and-rail sections, you should plan to set at least the corner

ROUND PENS

No matter what you call them—round pens, breaking pens or training pens—having one of these enclosures on your farm can be an asset in more ways than you might realize.

While the most obvious use of these pens is for starting young horses, the versatility of these areas as small, restricted work spaces and turnouts is great. They can be used in the following ways:

Two-year-old cutting horses are started in this pen, which is actually inside a very large indoor arena. Sold in kit form by a number of manufacturers, this useful type of pipe-and-plywood ring is common in the West and Northwest. The sections clamp together at top, center and bottom, allowing removal or expansion as needed. The standard 4'x 8' plywood sections bolt into place on the frame and prevent stray legs and feet from being placed outside the pen. Half of this covered arena is utility space for the round pen and adjacent tie rings for horses ready to be worked, and at the far end is a large square riding area. The solid wall of the arena is visible in the background with metal sheathing along its top to discourage chewing by tied horses. (Cutter Ridge Ranch, Livermore, California) Photo by N. W. Ambrosiano.

☑ To quarantine new horses or those who need separation from the rest of your stock

☑ For restricted turnout for horses recovering from illness or injury

☑ As smaller arenas for beginning riders (allows riders to build confidence without concerns over steering in large rings)

☑ As quiet working areas often of the same size as the 20-meter circle so common in dressage work

☑ As simple centers for ground work such as long-reining and longeing

☑ As breeding enclosures

There are many more variations on the round pen theme; several modern horse trainers with multi-disciplinary followings have built entire publications and video-tapes around the magic of a small, round pen. Regardless, putting together a pen with a diameter between 30' (for breaking young stock) and 66' or greater (for regular longeing and riding) can add surprising flexibility to your barn area.

For a sacking-out area, gentling young horses and keeping their busy minds focused on the trainer, smaller breaking pens are ideal, especially if you install a solid or mostly opaque wall that's at least 5' tall, preferably 7' or 8'. These become "training rooms," where quiet, steady progress can be made without distractions. Some trainers and traditional disciplines call for a heavy (at least 6" x 6") treated post, set deeply

Most round pens are, as the name implies, circular. Some, like this attractive one, are oblong instead, offering just enough room for limited riding as well as longeing. (Albemarle County Farm, Virginia) Photo by Lori Myers.

This simple round pen's openness and pleasant location make it a multi-use arena for its owners. Beginner riding lessons, starting babies and sacking out horses all happen here. (Horseshoe Farms, Diamond Beach, Florida) Photo by Lori Myers.

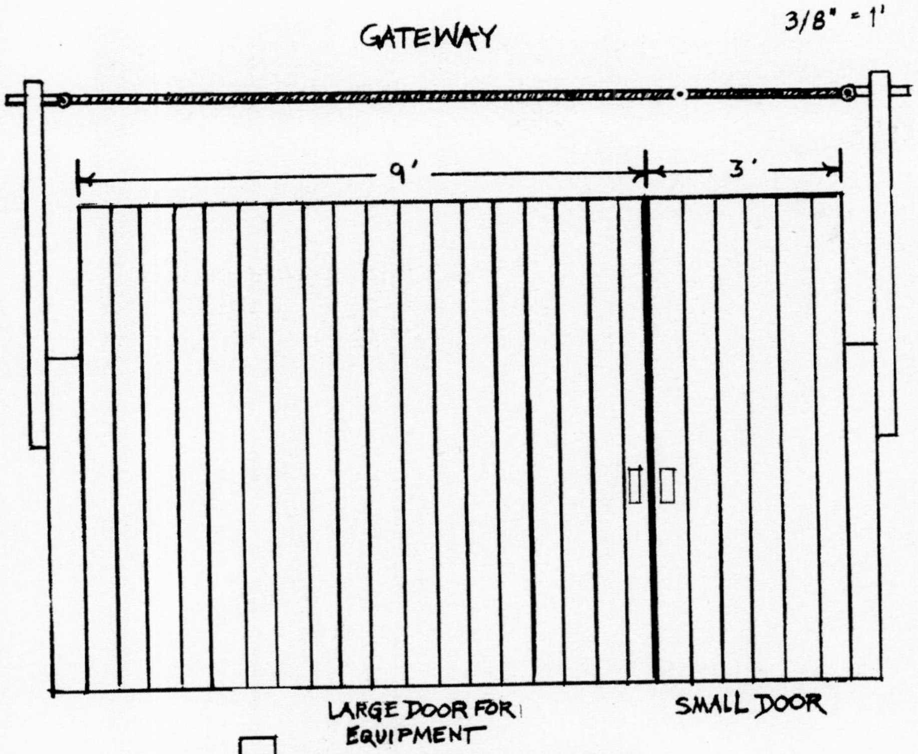

GATEWAY

3/8" = 1'

9'

3'

1 FT. SLANT OUTWARD

1'

1" x 4" TOP
BRACE BOARD

1" x 6" 8FT
PLANKS SPACED
ABOUT 2" APART

6" x 6" 8FT
BRACE POST

2" x 6" CENTER
BRACE BOARD
WITH CABLE ON
OUTSIDE TO ADD
STRENGTH

LARGE DOOR FOR
EQUIPMENT

SMALL DOOR

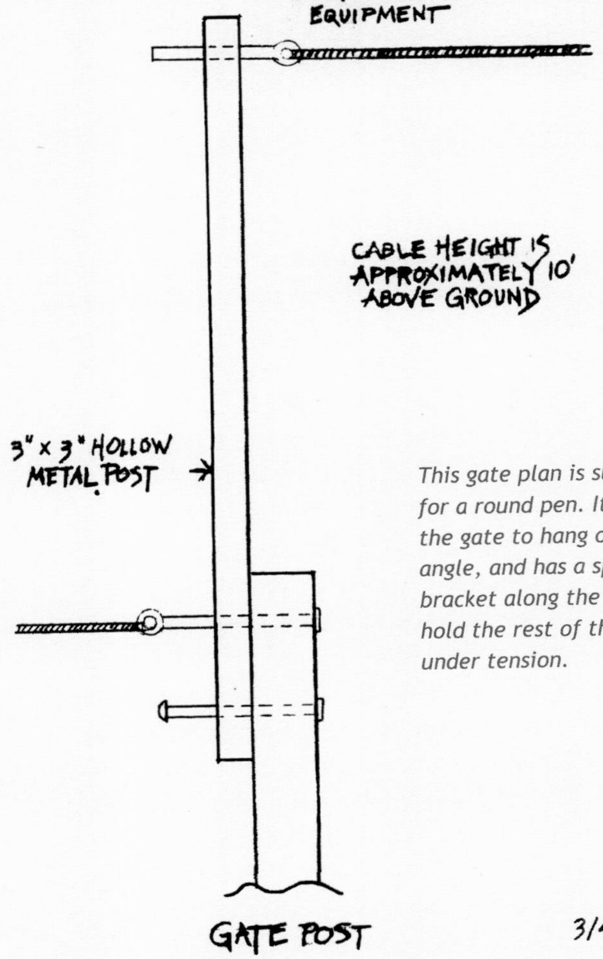

2" x 6" BOTTOM
BRACE BOARD

DIRT MOUNDED
ON OUTSIDE
TO BRACE
BOARD

CABLE HEIGHT IS
APPROXIMATELY 10'
ABOVE GROUND

2'6"

3" x 3" HOLLOW
METAL POST

*This gate plan is suitable
for a round pen. It allows
the gate to hang on an
angle, and has a special
bracket along the top to
hold the rest of the fencing
under tension.*

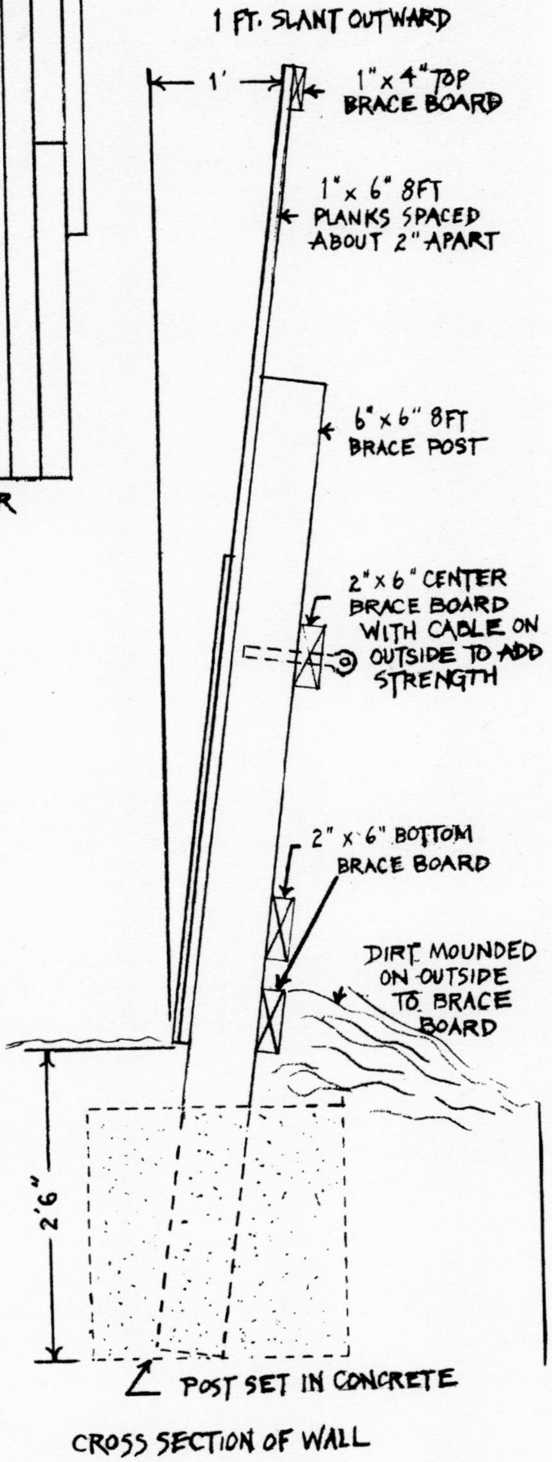

POST SET IN CONCRETE

GATE POST

3/4" = 1'

CROSS SECTION OF WALL

and reinforced, to be the center point of such a ring, good for tying up young-sters using a flexible stretch cord.

If starting young horses and retrain-ing rogues is on your list of duties, the breaking pen is ideal, especially if you also can install a 66' (20-meter) longe ring with regular fencing for the later stages of work.

If you plan only one small ring, the greatest flexibility comes with the 66' longe ring, provided your fence is at least 5' high and solid enough to with-stand the occasional collision or jump attempt from a loose or overly ener-getic horse. With a 66' or larger diam-eter ring, you can work horses reason-ably on a circle without excessive strain on their joints, although moderation is always called for depending on the ani-mal's condition and conformation.

Often built in a multipurpose or family facility, you'll find these small rings invaluable, especially if your bud-get is tight. You can install sand footing

or the ground covering of your choice for a fraction of the cost required for a full-sized arena. Thus, a tight-budget fa-cility can add at least one multi-weath-er working area without breaking the bank.

The variety of fencing used for round pens is a credit to human ingenuity. In areas with scrub trees, stacked saplings set between doubled posts provide an almost-solid wall, although their damp-weather deterioration reduces the pen's lifespan to less than 10 years. Traditional three-four- or five-board fence is attrac-tive, although too easily looked through for breaking pens. With these, you need the addition of more boards, plywood or other solid paneling at eye level.

Whether you use finished (smooth) boards or rough boards (which can save you money but add to your splin-ter collection), use screw or ring shank nails, sometimes called "pole barn nails," at least two to a plank, to ensure no boards pop loose during a rambunc-

Having built a really useful round pen, Bob Perks of LaGrange Farm decided to give it a roof, a free-standing pole barn construction that doesn't cut down on the fresh air by any means, but offers shade and shelter from falling rain. Photo by M. F. Harcourt.

One of the most traditional fencing types, this stacked-rail fence is constructed two panels at time, with paired posts that are wired together to secure the rails. Photo by M. F Harcourt

This is a pen that clearly gets plenty of use; the tools are near at hand. Leads, longe lines and side reins are all handy at the door. This pen was designed with a wide enough doorway for the tractor to enter for footing maintenance, but the doors are sized for human convenience: One is two-thirds size, left closed most of the time, and the other side is just wide enough for horse and trainer to enter safely. (LaGrange Farm, King George, Virginia) Photo by M. F. Harcourt.

sistant to chewing and kicking, and are the fastest way to provide a safe, round work area.

Using 4"x4" treated posts with heavy wire-mesh or chain link fence is a reasonable option, although as with any wire project, controlling loose wire ends and achieving a sag-free fit are priorities. Flexible plastic mesh is a possibility as well, but must be installed rigorously to avoid sagging sections.

Regardless of whether your fence is wire mesh, planks or panels, install the posts to the outside of the circle so that horses and riders don't accidentally catch knees, feet or elbows on the vertical supports.

To reduce overall collision potential, regardless of your fencing material, tilt the vertical supports from the ring's center by at least five degrees. If your soil does not hold posts well, consider packing the posts hard with rock and gravel in addition to the dirt. Concrete was once recommended for many fence-stabilizing projects, but rock, if available, allows drainage from wet weather so posts rot less quickly.

To add above-ground stability, run a cable around outside of the ring, from gate post to gate post the height of the top board, and tighten it with a buckle connector to keep the posts from falling further outwards as they age. Notches in the posts or staples will keep the cable from riding up and out of position. Support your gateposts against the cable's pull with either an angled brace post for each, or a lintel-type bar over the top, above the level of a rider's head.

The gate for your round pen doesn't need to be a large one, although if you plan to smooth the footing with a tractor and drag, you'll have to accommodate the vehicle's width. If no tractor work is on the agenda, a 4- or 5-foot gate is adequate. (Just be sure you buy

tious horse's turnout or work session. Be sure to check that warped boards don't create a space narrow enough to trap a misplaced hoof.

Those on a tighter budget often prefer pipe corrals, which are commercially available and easily (but not cheaply) shipped all over the country. Pipe sections, properly designed, can be dissembled and shifted to new locations, are re-

a good, wide rake for manually pulling the sand back to the working track after it's been kicked outward.) If you can make the gate of the same materials and appearance as the continuous line of fencing, you reduce the chance of a ring-sour horse fixating on it during work sessions. Be sure that the gate latches securely and can be reopened from both inside and outside the ring, for safety's sake. Always latch the gate securely during work sessions to prevent it from banging into a horse's body on a windy day, or giving way to a determined escape artist.

Unless your native soil is very sandy and rock-free, it's likely you'll need coarse sand, wood-chip products or other footing types to cushion the work area. Cushioning the full foot-print of the arena isn't essential since 90 percent of the work occurs along the rail, but life tends to be simpler if you do; your horse won't have to suddenly readjust to a different footing consistency when-

ever the circle is reduced or you ask for a change of direction.

Before installing footing, ensure that most of it stays in the ring by running a treated board "curb" around the base of the fence. Railroad ties are good for this, as are 2" x 8" treated boards, set just high enough to allow some water drainage below without too much footing washing out the same route. As you install this rim section, be sure your hardware was smoothly installed and is of the highest quality. This is the area where passing hooves have the greatest effect, easily knocking boards and light connectors loose and exposing fragile legs to laceration from detached nails, brackets or bolts.

Taking the vertical supports up to a 10' or better height and adding a roof, this round pen shows that you don't have to be out in the weather to longe. Photo by M. F. Harcourt.

Planning For Indoors

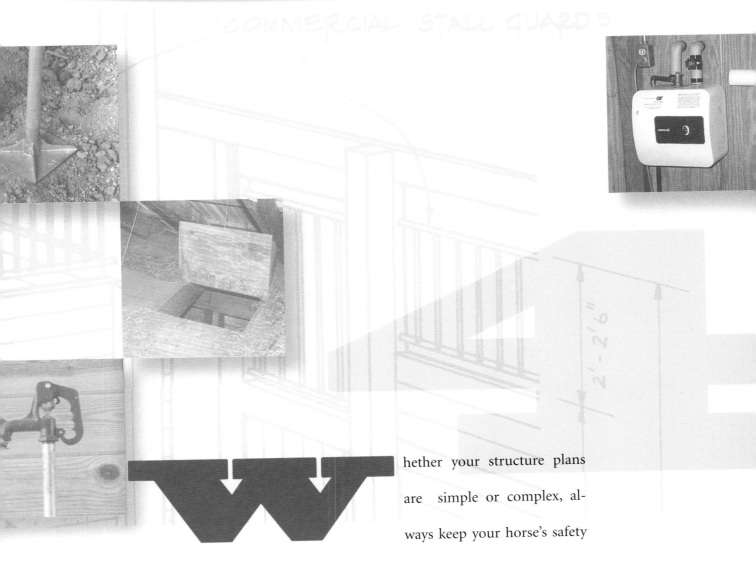

hether your structure plans are simple or complex, always keep your horse's safety in mind when designing the interior of your barn.

Many illnesses can be avoided with a well thought-out worming and vaccination program. Likewise, a well-designed facility can prevent injuries to horses and those working around them.

Managing for safety isn't difficult, but it does take thought and persistence. Be certain the right equipment is used in horse handling. Make sure the activities being conducted on the property are compatible with the facility design.

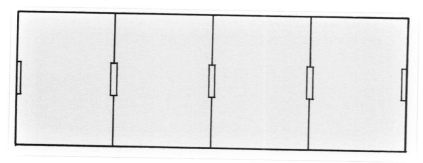

An illustration of how one might place fluorescent lights between stalls so each stall gets light from two directions. By hanging the lights over the partitions, there is no need to buy two lights per stall.

Design the interior layout as ergonomically as a gourmet chef would design her kitchen. Where will you need extra room for grooming, tacking up, shoeing, leg wrapping or general vet work? Where will your tack be in relation to your grooming area? How many extra steps can you save by making it easy to feed and care for your horses?

Easy labor should be your mantra to remind you to make design decisions that result in a safer, more convenient and less expensive facility. In the chapter that follows, you'll find some good ideas for doing just that.

LIGHTING

One of the things that can make a barn either a pleasure or a pain to work in is the lighting. For those interested in anything more than the most cursory of grooming and management chores, a well-lit view of the scene is essential.

Natural lighting not only makes a facility more comfortable and enjoyable for you but also for your animals. They will be more cooperative when handled if the surroundings are open and have adequate lighting. Horses have an intense natural fear of close, dark places. That is why it takes time to train them to go into trailers, and why they are hesitant about new areas where they cannot see well.

Skylights or windows in the roof can provide large amounts of natural light, and skylights that open have the added benefit of providing extra ventilation on demand. Skylights can range from the simplest of corrugated, fiberglass-reinforced plastic sheets, to small domes, to complex window packages.

If you choose to install clear panels, place them directly over key working areas, such as the main aisle in which you plan to groom, or over each stall, as the light will not be diffused very far from its source. If you have skylights installed by a professional, insist on a guarantee that they will not leak, as any perforation in the roof is likely to do so unless perfectly installed. Plenty of flashing and/or caulking is essential for a tight seal around each nail or screw and around the edges.

If piercing the roof is not to your liking, add a clear Plexiglas border at the top of the walls, just below the roof, to take advantage of low sunrays in the winter. This sky panel will be amply shaded in the hot summer when the sun's angle is more directly overhead. Adding this clear border all the way around dramatically increases your natural lighting, as long as you do not extend your roof's overhang too far out.

Even if you decide not to install lights inside the stalls, you'll need electricity for general lighting, running radios, clippers, water heaters and vacuums. Indoor lights will allow you extra working time during dark, cold winter months long after an early sunset. This is the time you'll need lighting for such chores as blanketing, bandaging tired legs and chipping ice from buckets.

You can install just about any type of lighting as long as you place the lights for maximum effect. Both incandescent and fluorescent lights are safe, and the horses won't care. If you choose to install one light per stall, place it, well-shielded, in the center of the ceiling, not in one corner or along a wall. A horse casts a large shadow, effectively blocking a large part of a stall when there is a side-mounted light.

For even better working conditions after sundown, install two lights, placed opposite one another along the walls, so you won't discover your horses halfway clipped or groomed in the clear light of morning. If you run your stall dividers up partway but not all the way to the ceiling, you can place fluorescent shop lights between each stall and get the benefit of two lights per stall, at a lesser cost. For example, in a four-stall barn, you'd have to install eight lights to have dual light sources in each. If you instead use partial dividers, five lights can be split between the stalls and still give you light from two sides in each. Also, choose low temperature ballasts for cold climate use.

Having one area extremely well lit for routine close work or emergency vet work is very handy, as most emergencies seem to occur after dark. If you need to keep a horse well groomed or prepare one for competition or other activities, having a well-lit area in which to prepare will be useful. If you need to set up a foaling stall or wash horses during cold weather, you may want to use heat lamps. Plan your wiring so it will be suitable for carrying the required amount of current. Both 110 and 220 come in handy in at least one outlet in the work area.

While lights may seem to be the only electrical need in the barn, consider the number of appliances you may wish to plug in from time to time, such as horse clippers and vacuums. An outlet of the exterior type, with a spring-loaded cov-

er nearby or in each stall, can be useful, as are outlets around the tack room and general work areas. Other potential uses for electricity in your facility include bathroom heaters, hot water tanks, washers and dryers, intercoms, video cameras, and security systems. In fact, just about anything you can imagine could be added to your facility, depending on what you want and how much you want to spend. The key is to put in as much electrical capability as your budget allows, and then some.

All in all, your minimum needs include:

 One light every 10' of aisle way

 One double electrical outlet at each end of the barn

 One light in each stall

 One well-lit area for grooming and vet work

As for the outside, a floodlight over one or more barn doors is a lifesaver if you feed after dark or unload after a show. You can install a simple floodlight with an easily located switch or motion detector, or look into getting a light-activated fixture, such as a street lamp that will brighten the area as soon as the natural outdoor lighting drops below a certain level.

When installing electrical fixtures, do it right the first time. Unless you really know what you are doing, you'll have more peace of mind if the wiring and electrical outlets have been safely and correctly installed by a professional. This will reduce the chances of a barn fire. Be sure to enclose all lines, both interior and exterior, in metal conduit, protecting the wires from the predations of rodents, horses and sharp objects. Horses are hopelessly curious about wires crossing their walls, especially foals and horses on stall rest, and

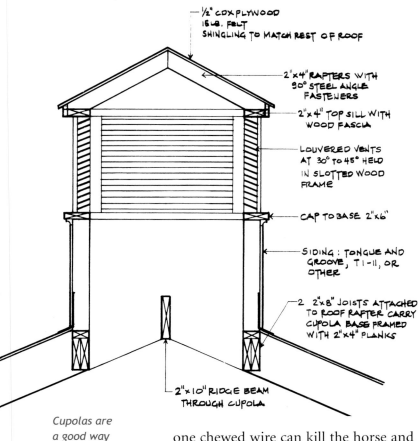

½" COX PLYWOOD
16LB. FELT
SHINGLING TO MATCH REST OF ROOF

2"×4" RAFTERS WITH
90° STEEL ANGLE
FASTENERS

2"×4" TOP SILL WITH
WOOD FASCIA

LOUVERED VENTS
AT 30° TO 45° HELD
IN SLOTTED WOOD
FRAME

CAP TO BASE 2"×6"

SIDING : TONGUE AND
GROOVE, T1-11, OR
OTHER

2 2"×8" JOISTS ATTACHED
TO ROOF RAFTER CARRY
CUPOLA BASE FRAMED
WITH 2"×4" PLANKS

2"×10" RIDGE BEAM
THROUGH CUPOLA

Cupolas are a good way to provide ventilation.

one chewed wire can kill the horse and burn down the barn.

Install a switchbox that allows for ample future expansion. Adding sheds, stalls and work areas will be less of a nuisance if lighting is prepared ahead of time. Inside the barn, recess all light fixtures, if possible. If not, be sure to cover them with a sturdy grate or screening to protect them from breakage and possible electrocution of either people or animals. Place them as high as possible; you may not think your fixtures are low until one is broken by a rearing horse, or a rake or shovel being used in its vicinity.

VENTILATION

Adequate ventilation in your facility means providing a clean, fresh exchange of air in your barn. By that we mean fresh air, not drafts. Loose construction, with no means of shifting or shutting off the breeze, can aggravate a horse with respiratory ill-

ness to the point of pneumonia. Good, planned air flow is necessary not only for your horse's health but also for yours and that of any employees you may have.

Natural ventilation can be used in most cases by laying out the structure to take advantage of the prevailing winds, correctly spacing doors and windows, and adding any cupolas or vents you may feel are necessary for your climate. A knowledgeable neighbor or the county extension office can give you tips on how the extremes of climate and prevailing wind patterns affect horses in your area. They can also offer suggestions for aligning your facility most sensibly on your property by taking climate into account.

If you can manage good air flow without extra fans, so much the better. Some steps can go a long way, such as adding 6- to 8" spacing along the eaves, using a cupola, or inserting ridge vents the entire length of the building. You want the hot air to rise. Pulling in cool air on the floor level and spreading a cool breeze along the way is essential in hot, muggy weather.

If you're thinking of a handsome cupola as an addition to your barn's profile, a rule of thumb for dimensions would be 1" of cupola length for every foot of roof length. It's easy to underestimate the size of one when you're looking from the ground, but notice that on a four stall, 48' barn for example, a well-sized cupola is about the size of a stall door.

It's not critical that your cupola be the ventilating type. You may simply use it as a dove cote or pure ornament. But by piercing the roof and putting in an exhaust fan, you can draw a remarkable amount of hot, stale air out of your building, allowing fresh air to pour in at lower levels. If you live in a climate that is cold in the winter, be sure to provide

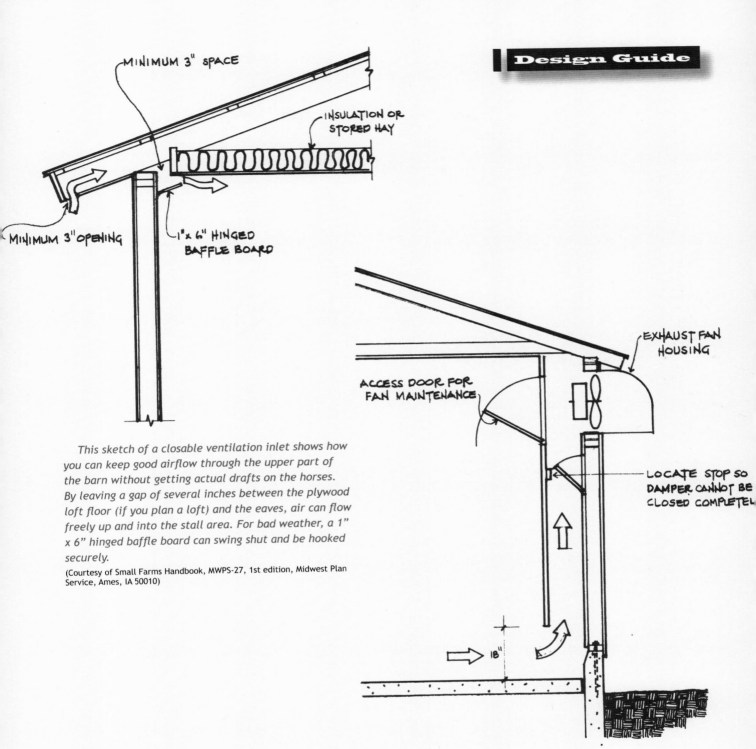

MINIMUM 3" SPACE

INSULATION OR STORED HAY

MINIMUM 3" OPENING

1" x 6" HINGED BAFFLE BOARD

This sketch of a closable ventilation inlet shows how you can keep good airflow through the upper part of the barn without getting actual drafts on the horses. By leaving a gap of several inches between the plywood loft floor (if you plan a loft) and the eaves, air can flow freely up and into the stall area. For bad weather, a 1" x 6" hinged baffle board can swing shut and be hooked securely.

(Courtesy of Small Farms Handbook, MWPS-27, 1st edition, Midwest Plan Service, Ames, IA 50010)

EXHAUST FAN HOUSING

ACCESS DOOR FOR FAN MAINTENANCE

LOCATE STOP SO DAMPER CANNOT BE CLOSED COMPLETEL

18"

If you want to keep the barn warm, but at the same time want good, constant airflow, try this simple installation. A double wall, with the fan set in the outside wall, is built at one end of the barn, and an 18" gap is left at the bottom of the inner wall. Cool air flows across the floor and out, drawn by the fan, while your warm air circulates snugly indoors. A damper, like that in a fireplace, controls the rate of airflow, but has a block to prevent complete closure in case the fan is still running when the damper closes. Be sure and leave a flapped door for maintenance access to the fan, as shown.

(Courtesy of Small Farms Handbook, MWPS-27, 1st edition, 1984, Midwest Plan Service, Ames, IA 50010)

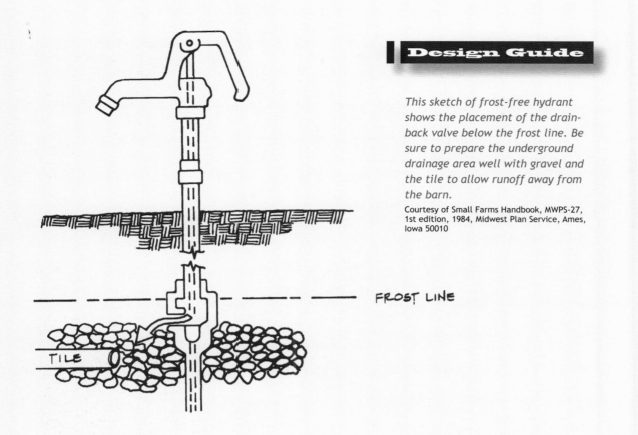

This sketch of frost-free hydrant shows the placement of the drain-back valve below the frost line. Be sure to prepare the underground drainage area well with gravel and the tile to allow runoff away from the barn.

Courtesy of Small Farms Handbook, MWPS-27, 1st edition, 1984, Midwest Plan Service, Ames, Iowa 50010

FROST LINE

TILE

some way to control the air flow during winter weather.

In particularly close, hot climates, vents along the stall wall can help. Placed about 2' above the floor, a long narrow slot vent, 2" by 7' can keep the stall floor much healthier, especially for foals who don't benefit from air flow only at window level. Adding skylights that can be raised also allows heat to escape.

If you need to keep many horses in a small area, be aware that each animal puts out two gallons of moisture daily by breathing, plus a considerable amount of body heat and airborne contaminants. Research has shown that horses' lungs are noticeably damaged by the effects of dust, mold, ammonia and other pollutants in the air. Owners of performance horses and breeding stock should be especially aware of this, as a horse's performance suffers greatly from diminished lung capacity. Increased sickness in a particular barn can also be attributed to bad ventilation.

To clear the air with more than passive measures, you can install fan and venting systems large or small. Three to five air exchanges per hour will usually provide adequate ventilation. You may only need to place attic fans in the eaves above the stalls, or you could tackle the problem the way the professional agricultural folks do. They have access to good information and plans for adding mechanical ventilation systems, often at surprisingly reasonable prices. Here again, your County Extension Agent can offer suggestions, many of which are simple enough for almost anyone to incorporate into existing building plans.

WATER

Two main factors will affect your plumbing plans in the barn: your water supply and what you hook up to it. Everything you do in and around your facility will be easier if your

Automatic stall waterer. (Private farm, Southern Pines, North Carolina) Photo by M. F. Harcourt

This recessed Murdock hydrant offers safety and durability. Note the drain area under the bucket which ties into the drain system for the adjoining wash stall and a nearby courtyard. (Albemarle County, Virginia, barn) Photo by M. F. Harcourt

water source is both adequate in supply and efficiently laid out.

The purpose of having a water supply is to provide water in adequate amounts to maintain your horses' good health. Horses require a large quantity of water—5 to 15 gallons of it—on a daily basis. Major health problems in horses result from a lack of water for over just a few hours. Climatic conditions as well as exercise and carrying a foal can increase the normal amount a horse consumes, so providing water a horse can drink whenever he's thirsty, day and night, is a necessity.

Hot weather increases a horse's consumption of water, just as it does in people, but cold weather also puts an increased demand on your horse's need for water. Winter feeds are usually drier than summer grasses, and a horse needs even greater amounts of water to compensate for dry feedstuffs.

The size of your water supply depends on the size of your operation, the number of animals you care for, and the type of operation you have. The simpler your operation (a run-in shed for a pleasure horse, for example), the simpler your water needs. Of course, if you

add the requirements of horse bathing and blanket washing, the amounts increase proportionally.

First, decide if you need to dig a second well or if your existing well will support both your household and your barn. If you are on a city or county water line, check to see about an increase in cost if you use large volumes of water. This may have bearing on whether you install a barn-only well or merely practice great efficiency in stable chores that require water. Finally, if you drill a well

The main water tank and heating unit are major components in the gravity-fill watering system. A hinged wooded lid coated with spar varnish keeps the tank from being contaminated or a danger to farm creatures. Photos by M. F. Harcourt

The main tank with the cover lifted shows the floater valve with shutoff coming from the heating unit, upper right. Below the surface is the open end of the main pipe leading to the buckets. At this end of the pipe, a shutoff valve allows the system to be sealed for drainage from the other end.

This frost-free hydrant has a drain-back valve below the frost line. Ken & Helen Montgomery, Aberdeen, North Carolina

The gravity-flow "bucket" is shown in its insulated corner box. Notice that there are no protruding objects for a horse to break or hurt himself on. A simple interior frame holds the front of tongue-in-groove 1" x 6"s with a piece of plywood forming the surface through which the PVC pipe protrudes. The interior of the box is filled with non-toxic foam insulation.

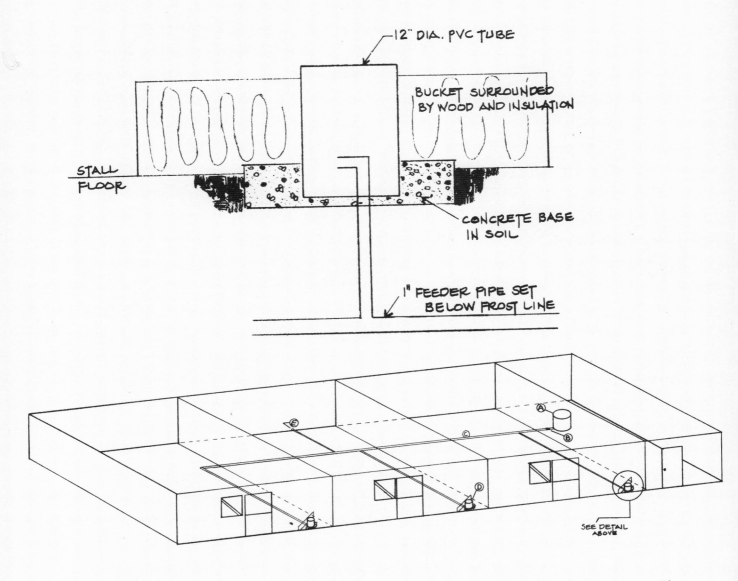

12" DIA. PVC TUBE

BUCKET SURROUNDED BY WOOD AND INSULATION

STALL FLOOR

CONCRETE BASE IN SOIL

1" FEEDER PIPE SET BELOW FROST LINE

SEE DETAIL ABOVE

This watering system allows for gravity flow to each bucket and is frost-free as well. A heated, main holding tank in the tack room, kept filled by a simple toilet tank-type floater valve, is the heart of the system. From there, PVC pipe runs below the frost line the length of the barn, branching to each stall's corner waterer.

The "buckets" are actually short sections of 12" PVC pipe set into a concrete pad through which the feeder pipe runs. Each is set inside an insulated corner box and is leveled to the height of the main tank's water level. No part of the feeder pipe is ever exposed to air or horses' teeth.

To drain the system, a separate drain pipe and valve branch off the main pipe. With the closure of the main tank's valve and the drain valve's opening, all the waterers can be flushed out at once. Before flushing the system, one should be sure to pick out extra debris from the buckets so that it doesn't block the pipes on the way out. A quick swish around the inside with a scrub brush cleans off the sides. If you have a wet/dry vacuum, you can suck the remnants out of each waterer in moments.

The model shown was designed by Roy Rottenberry and Grace Dawson, and is installed in the barn of Mrs. Dawson and her daughter Phyllis at Windchase Farm in Hillsborough, Virginia. While this design is "homemade," other designs can be purchased from some farm supply centers.

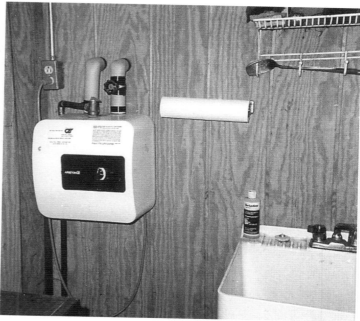

Simple and extremely long-lasting, fiber-reinforced rubber corner buckets serve well as waterers. Such a bucket is especially solid when secured by a corner bracket to lock the handle in place and a metal lip hooked under the rim of the bucket. The rubber material won't produce sharp edges if kicked or stomped, and during the winter, ice can be popped off the water's surface by a thump on the flexible bucket's side. Photo by N.W. Ambrosiano

This Ariston 4-gallon water heater operates off 110 current, making it easy to install in a barn for tack cleaning, adding to bran mashes and spot cleaning manure stains on horses. The small quantity of water stored is not enough for a complete horse's hot bath, but for many other needs it's ideal. Beside it a deep utility sink and work shelf make tack cleaning simple, and hanging wire racks prevent sponges and wet brushes from turning rank. (Chestnut Hill, Spring Grove, Virginia). Photo by N. W. Ambrosiano

for the barn, be sure to have the water tested for chemical and bacterial contamination.

Plumbing Installations

When planning plumbing installations, one of the decisions you will have to make is whether you want automatic waterers. They are convenient and can save you time and effort in keeping an adequate supply of water in front of your horse. They come with or without heaters, and the better models are splash- and spill-proof, effectively defeating the efforts of bored horses to flood their stalls.

Like buckets, automatic waterers need cleaning on a regular basis. And they are not without drawbacks: Most importantly, you cannot keep track of the amount of water your horse con-

sumes on a daily basis. By mentally recording the daily consumption of water, you can catch colic and other digestive problems before they become serious.

Automatic waterers can also be broken, usually in the middle of the night, by an errant kick from your horse and thus flood the barn. In chilly climates, even heated ones can freeze, as the water lines to them are always kept on. Use metal pipes to supply waterers so you can thaw frozen sections with a blow torch. Metal supply lines also prevent horse teeth from opening sections unexpectedly. To minimize frozen pipes in hard-to-reach areas, wrap supply lines with electric heater tape in those sections beyond your horses' reach.

If you plan to use faucets as your water source, drain back valves are the most reliable type to use. They will give you a long period of service with little or no maintenance and will minimize those annoying mud puddles that form around water supplies. These valves automatically release water into a small below-ground drain field when the faucet is turned off, instead of letting water lie in the pipe near the faucet. That way, no water sits in the pipe above-ground, where it may freeze. Drain-back valves can be installed in several locations in the barn and in pastures to provide water under the most severe weather conditions. Just remember to check the freeze line in your area (depths range from 0 to 2' or 4' underground).

If you need hot and cold running water, include a hot water tank in your plans as well as suitable plumbing to handle the additional line. This is not complicated, and it is one measure that will make much of your horse keeping easier. The small tank is relatively inexpensive to buy and run, and if it is a quick-recovery type, it can provide plenty of hot water for a minimal investment. This will allow you to add a washer and dryer to your barn to keep all the towels, wraps, blankets and other horse items clean while avoiding wear and tear on your household appliances.

Other water users, such as bathrooms and wash pits, are wonderful additions to a barn, but not essential. If you think you might want to install these later, arrange the plumbing so that fixtures and associated structures can be added. Just be sure that the runoff from these will not overload your current septic system. Before you decide on an additional bathroom, check your county regulations, zoning permit and percolation permit to be sure it is allowed.

Just as with electrical wiring, put in as much plumbing as your budget will stand. Even if you don't add the bathroom, washer and dryer or wash pit right away, install the line and connections. Remember to keep diagrams of the lines, valves and fixtures for future maintenance reference.

Obviously, you don't need all this fancy stuff to enjoy your horse, but if you are into showing, eventing, hunting or training, these things make it easier and quicker to keep your equipment and horses clean and well-maintained.

FLOORING

Before you finish your stalls and fill them with bedding, remember they will take a great deal of abuse from the combination of your horse's feet and the waste products of so large an animal. If the floor isn't tough, you'll soon have a bog in the barn, no matter how deep the bedding.

The flooring you choose needs to provide good drainage, easy manure removal, and minimal dust and noise. While it's hard to find all that in one affordable surface, you have various options: mats over concrete, packed earth, and even straight sand may fit your needs.

It's a matter of considering what you have available and how you'd like to arrange your stalls.

Drainage in stalls or a shed can be helped by laying a mini-drain field in each stall or by adding a mini-drain field running through the center of your stalls. To do this, dig a hole about 3' in diameter in the center of the stall. Make it deep enough to reach a layer of dirt that will drain adequately. Then fill the hole with large-diameter gravel, or alternate layers of sand and gravel, and tamp that firmly into place before covering it with the dirt you have selected

or the shed or stall. Another good, simple way to prepare a stall is to lay a 6- to 12" deep layer of fine gravel or stone dust, and then place your dirt, clay or other materials over it.

If you find a layer of bedrock below your top soil, you can use it as the "bottom" of a mini-drain field by channeling urine out of the barn and allowing it to drain down and away from the facility. Not that much is really going to drain, as the bedding will absorb much of the moisture overnight.

Porous "popcorn" asphalt is another flooring that works fairly well. It provides a solid covering for the drain field over which you can spread a good layer of bedding to protect a horse's tender joints from the hardness and roughness of the surface. To get a good surface for horses, have the hot asphalt raked rather than hot-rolled during installation, and when you order it, be sure to ask for the type with large particles. When you order the asphalt, request one that is a urine-resistant emulsion of double tar with little to no sand.

Another fairly costly but well-supported flooring method is to place a grid of pressure-treated 2x4's edgewise across the stall over a 12" base of gravel and fill stone dust between the boards. The boards should be between 1 ½ and 4" apart, braced at either end. The stone dust should be packed firmly between them to the top of the board edges. This will drain superbly, providing better traction and more give than a concrete floor. It will also allow those cleaning the stall to use a shovel, which can be slid along the boards for the last part of the cleaning. It takes 45 12' 2x 4's to grid a 12' x 12' stall, and it is more expensive than some matting systems. If your are lucky enough to find inexpensive 2x4's, or can get rough cut, untreated oak, you may be able to reduce the cost or at least bring it in line with stall mats. But for horses who paw at stall doors, it may solve a management problem and be worth the extra cost and effort to install.

A simpler drainage option involves packing the flooring soil in a gentle

The aisle of this barn is kept safe and clear because of its width, properly installed doors and equipment storage. (Private farm, Southern Pines, North Carolina). Photo by M.F. Harcourt

slope to one corner or to the rear of the stall. Clean the damp area thoroughly and lime it weekly to prevent odors and bacterial buildup. When grading the flooring, keep the slope to a maximum of 3 degrees so your horse's legs will not become strained by standing at such an angle for long periods of time.

You have another choice in the final layer of dirt that lines the stall. Classic flooring is hard-packed clay, which offers a tough, yet forgiving surface that won't scrape your horses or cause their legs to swell if they must stand on it for many hours. Thick clay is the best, packed to a nearly hoof-proof surface, because it lets urine run off to a drain hole and only needs to be reworked once a year or so. Some people order baseball diamond clay for this and pack it no less than 4" deep. If no clay is available, any thick, packable soil free of stones and sandy portions (which disintegrate) will do, but it may require more frequent repacking.

To form the most solid soil flooring, you need a heavy pounding tool with a flat foot and a waist-high handle you can raise and drop repeatedly while moving slowly around the stall to level

Use a tamping tool, as shown, to pack down your floor soil adequately. Be sure you work the floor thoroughly from one side to the other. Photo by N.W. Ambrosiano

the entire surface. Motor-driven settlers for concrete do the best job, provided you have both a hose and a supply of loose clay on hand to alter the dampness or dryness of the surface as you go along.

If you plan to wash and disinfect your stalls, as in a breeding or hospital arrangement, more efficient drainage is in order. In some operations, actual drains are placed in the corners or center of the

Interlocking cushioned bricks provide a quiet, non-slip surface for aisles and passages. Photo by N. W. Ambrosiano

stalls, and solid flooring such as concrete is covered with commercially sold rubber stall mats. Most farm managers lay bedding over the mats, removing it when the stall is disinfected and replacing it with clean straw or wood chips before the horses are brought back in.

Others prefer not to place bedding over the mats at all, washing the stall contents into the drains daily. The drains empty into a sluicing system that can be flushed weekly, and thus the stalls are kept clean and fairly odor-free. Since the only waste material is pure manure and not bedding, stall cleaning is easier, and there is less of a manure pile to be disposed of. The pure manure can be sold or given away as organic compost.

The drawback to using matting in your stalls with no additional bedding is that in cold climates, the facility may be colder than you like. Also, it's hard to beat the aesthetics of a barn well bedded with gleaming straw or well-banked wood shavings or sawdust.

The stall floor mat industry has expanded extensively in recent years, and ribbed, perforated, smooth and pebbled mats are available. Excellent plastic grid systems that allow drainage into a prepared stall base are also for sale. In all cases, the base must be firm, and very smooth and flat from one side of the stall to another. Otherwise, any matting you install will shift, warp and possibly even catch a hoof or shoe in a gap at a seam, at possible harm to the horse.

Your mat choice will depend on the type of horse keeping you do, as full-time stalled horses, older animals, and mares with foals may produce far more urine than other horses. Both your mats and your stall base must be configured to accommodate that flow.

If your stall base and local environmental arrangements permit urine to drain safely beneath the stall, purifying itself as it percolates through soil, sand and gravel layers, a mat choice that is either perforated or a simple grid is a good idea. It will reduce your monthly bedding costs, leaving less dampness for the bedding to soak up. The daily or weekly use of hydrated lime or another commercial stall freshener will neutralize any remaining ammonia smell.

Another less-publicized product that is good in a draining stall is something called "cow carpet"—a woven synthetic "rug" that is sold in rolls, and like rubber mats, prevents the stall base from being dug up into the bedding during normal use.

If you prefer to have all the moisture carried physically from the stall in the bedding, solid mats (with ridges or pebbling to prevent slips) are effective, but you will see an increase in the amount of bedding used, with an accompanying increase in your financial burden.

Check the advertisements in national and regional horse publications for mat products and availability in your area, and watch for product-review articles in magazines and newsletters.

STALL DESIGN

Stall Size

Based on the size of the horse or pony you are housing, you have some options in stall size. The smallest comfortable stall for an average horse is 10' x 10', which allows just enough room for the animal to turn around in and lie down safely. A 12' area is more generous for the animal and gives you more working space if you choose to do your grooming in the stall. However, consider that in enlarging your stalls, you increase your materials—bedding, roof area, everything—by 44 percent or more. It's a simple calculation of area: a 10' x 10' stall is 100 square feet. A 12' x 12' stall is 144 square feet, needing almost half again as much bedding, roofing and so on.

Owners of stallions or broodmares really have no choice, as breeding operations must allow space for mares to foal, and stallions to live comfortably. They need stalls that are 12' x 12', 14' x 14' or larger.

A simple way to construct a foaling stall on short notice is to remove a partition between two regular stalls. Removable partitions can be made by sliding 1" x 6" or 2" x 6" boards down between a channel of boards (2" x 2"s or 4" x 4"s are fine) on opposite walls. Convertible stalls are also useful when you have a horse on extended stall rest.

Constructing all your stalls as 12' x 12' with removable partitions between each allows your facility a wide range of uses and isn't that hard to do in the initial construction stages.

If you have planned a hayloft (not recommended as hay stored in a shed is safer from fire hazards, easier to use and won't fall through cracks onto your horses' coats and into their lungs), allow enough clearance so that no part of the supporting structure is lower than 10'. That way, your horses will not injure themselves when they throw their heads in excitement or fear. If you are working on a tight budget and will have smaller horses in the barn, 8' ceilings are acceptable, but be sure to completely recess your light fixtures.

Doors

Standard stall doorways should be 4' wide. Once the doorway is framed, however, you have choices as to filler. Light or heavy doors, screens, chains or web barriers are all appropriate, depending on your animals and the time they spend indoors. If you wish to save expenses during construction, consider using inexpensive stall guards as a start-

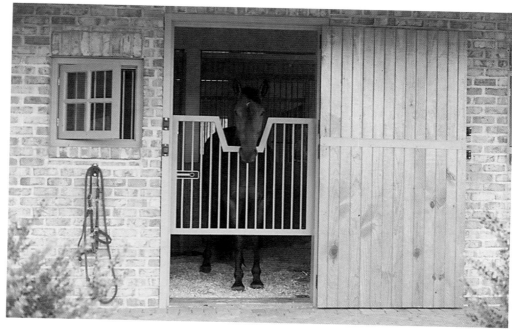

up door. Later, you can add either pre-constructed doors or doors you make yourself.

Solid swinging doors can either be the Dutch types (separate upper and lower parts) or a full door, which may be solid or have steel mesh or bars on the upper portion for light and ventilation. You can make standard Dutch doors fairly simply, provided you measure each section carefully and conduct several trial installations as you go along. The trick is to have both doors hang securely and evenly so they do not swing open or closed on their own, striking the horse unexpectedly. Doors must never swing into the stall, as they can crush the horse against the frame as he passes or become stuck in the bedding as they swing. If your Dutch doors open to the outside, be sure the latches are outside as well. In case of fire, you can then free the horses without entering the barn. Hang a halter outside as well to make rescues easier still.

A sliding door may be more expensive initially but it adds aisle space to your facility and is generally safer. It's much harder, for example, for a horse to get a hip hung on a sliding door than on one that swings into his path as he passes.

Doors can also be purchased as complete, ready-to-hang units. Most are sold in standard 4' widths, whether they swing or slide. Search horse magazines for the many companies advertising these and other barn materials, but

To build a traditional Dutch door, you need simple, solid materials such as 2" x 6"s, 5/8-inch plywood, heavy strap or Tee hinges and metal edging for sections vulnerable to chewing. While this door plan shows a full 4-foot door, if you are building with 4" x 4" posts set four feet on center, you will need to make this door 44 inches wide, not 48, so it will fit neatly between the posts.

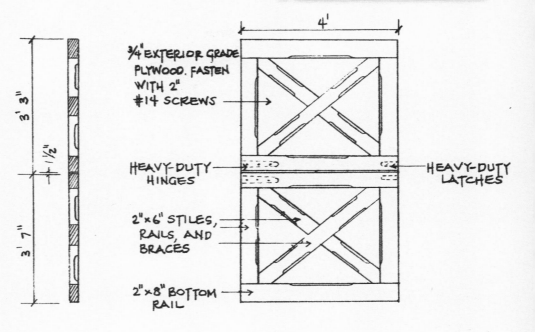

3' 3"

1½"

3' 7"

4'

¾" EXTERIOR GRADE PLYWOOD. FASTEN WITH 2" #14 SCREWS

HEAVY-DUTY HINGES

HEAVY-DUTY LATCHES

2"x6" STILES, RAILS, AND BRACES

2"x8" BOTTOM RAIL

Sample stall specifications show a sliding door installed with treated lumber near the floors, commercial stall guards, and partitions made of planks set between 2" x 2" channels on the walls. (Small Farms Handbook, MWPS-27, 1st edition, 1984, Midwest Plan Service, Ames, IA 50010)

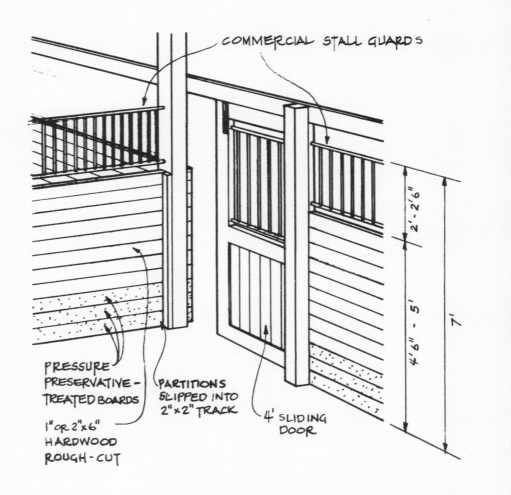

COMMERCIAL STALL GUARDS

2'-2'6"

4'6" - 5'

7'

PRESSURE PRESERVATIVE-TREATED BOARDS

1" or 2"x6" HARDWOOD ROUGH-CUT

PARTITIONS SLIPPED INTO 2"x2" TRACK

4' SLIDING DOOR

If the shape of your building precludes a large sliding door, it's not impossible to plan a hinged one that stays solidly in place. This extremely large swinging aisle door is attached using a piano-type hinge from top to bottom, and it latches against a stop post when open. (Albemarle County). Photo by M. F. Harcourt

Close-up showing piano hinge of large swinging door, which bolts into a 10" x 10" post set in concrete. Also visible is a fire extinguisher; one or more of which should be close at hand in every barn. Check with local authorities on required number and spacing of fire extinguishers. Photo by M. F. Harcourt

beware that shipping charges add substantially to the expense. In assessing whether to build or buy doors, consider purchase and shipping price versus your own time and effort combined with the cost of the materials.

Heavy gratings are a good option when hung with or instead of standard doors. They come in squares a horse can look over, tall rectangles that keep horses from biting passersby, or rectangles with a scooped opening. They are readily purchased and easily installed on eyebolts or even as sliders. Again, be sure to place them for safety so they open to the outside of the stall.

The solidity of what you install is important. Due to its frequent daily use, a door takes more abuse than the rest of the barn structure. Where lighter lumber may be adequate along a partition, use no less than ¾" plywood and 2" x 6" boards for the door's framing and backing. A

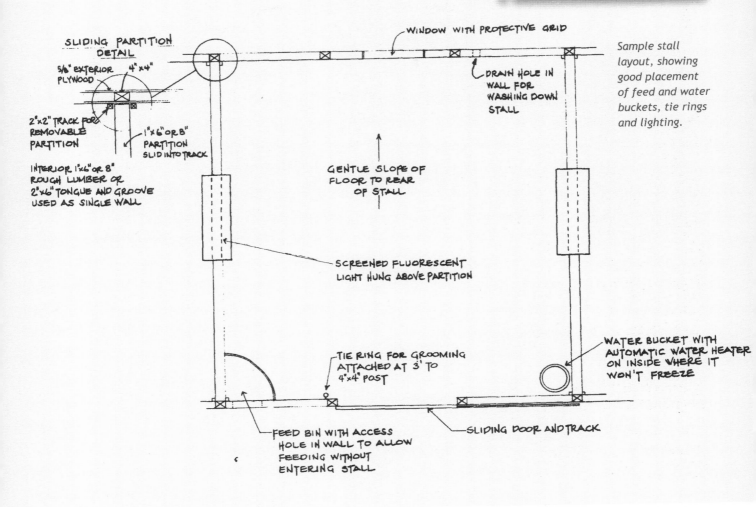

SLIDING PARTITION DETAIL

5/8" EXTERIOR PLYWOOD

4"x4"

2"x2" TRACK FOR REMOVABLE PARTITION

1"x6" OR 8" PARTITION SLID INTO TRACK

INTERIOR 1"x6" OR 8" ROUGH LUMBER OR 2"x6" TONGUE AND GROOVE USED AS SINGLE WALL

WINDOW WITH PROTECTIVE GRID

DRAIN HOLE IN WALL FOR WASHING DOWN STALL

GENTLE SLOPE OF FLOOR TO REAR OF STALL

SCREENED FLUORESCENT LIGHT HUNG ABOVE PARTITION

TIE RING FOR GROOMING ATTACHED AT 3' TO 4"x4" POST

WATER BUCKET WITH AUTOMATIC WATER HEATER ON INSIDE WHERE IT WON'T FREEZE

FEED BIN WITH ACCESS HOLE IN WALL TO ALLOW FEEDING WITHOUT ENTERING STALL

SLIDING DOOR AND TRACK

Sample stall layout, showing good placement of feed and water buckets, tie rings and lighting.

frustrated horse, either pawing or plunging against a door, can test the strength of your construction to its limits. This is especially true if you have a boarding operation or a breeding facility.

Along the tops of the doors, as on every possible chewing surface, place a thin sheet of metal to discourage destructive teeth. Throughout the stall, one hopes to have few places where a chewing or cribbing horse can get a grip, and door and partition tops are inevitable targets.

To foil equine teeth, buy drywall corner beading at a builder's supply store. It's very cheap, tooth-proof and has holes predrilled so it can be screwed to the door. Alternately, buy 4- or 6" wide rolls of aluminum flashing at a hardware store, and tack it down with roofing tacks or other wide-headed nails. You need to unroll a length of metal exactly the size of the surface it will cover, cut it with tin snips, center it on the surface, and firmly bend down the metal on either side of the door or partition top. It should form a three-sided channel that is smooth and clean-looking, with no untrimmed edges to snag passersby. As it wears, remember to keep an eye out for loose tacks that can catch a passing horse or fall out only to be stepped on. A piece of form-fitted galvanized steel can also be used.

The hardware on which to hang the door is almost more important than the door itself; if the hinges sag or tear

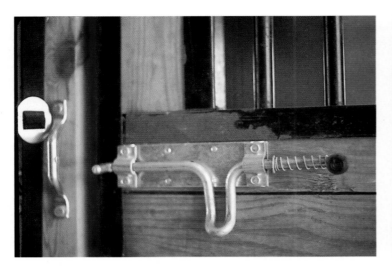

Stall latches that are spring loaded ensure that the latch bounces back flush with the door edge. This safety latch keeps your horse from catching its shoulder, side or hip should it get close to the door edge as it exits. Photo by M. F. Harcourt

Treated lumber at the bottom of stall partitions prolong partition life. (Ken & Helen Montgomery, Aberdeen, North Carolina). Photo by M. F. Harcourt

Where two sliding doors meet they can be latched together with a hook and eye, but that won't keep them from sliding as a unit once they're latched. Either install another hook and eye at the rear of one or both doors to hold the joined doors in place across the doorway, or get an adjustable hook like this one. Placed as shown, it can be adjusted for changes in the door's placement and also locked into position. (Ken & Helen Montgomery, Aberdeen, North Carolina). Photo by M. F. Harcourt

free, the door will be impossible to open or close safely and easily. Use the largest strap hinges, whether plain or ornamental, that can support a nearly 100-pound door for the best wear. Use two hinges per door on Dutch doors and three on full doors. The latch should be horse-proof, withstanding the efforts of nearly prehensile horse lips to open it and escape. A determined horse can easily lift a chain over a nail, flip a hook, or untie a rope. If you install the type of stall bolt with a handle that you raise and slide to free the bolt, be sure to put a screw eye through the center of the handle and attach a heavy snap to it so the horse cannot work a lip under the handle and lift it.

If your horse is prone to pawing at the door, a kick latch at the bottom as well as a top bolt will help. You can order toe-operated latches through farm or stable supply catalogs or national builders hardware supply.

If your barn has a center aisle, doors to the main entrance give you the option of increased climate control. Here again, you must decide if swinging doors or sliding doors fit better with your overall facility plan. Swinging doors require more space to open, and can bang or flap themselves to pieces in big winds if not securely fastened. Also, doors large enough to cover the opening of your barn often require maintenance as they age due to both their extreme weight pulling on the hinging apparatus and the effects of the wind. If you use swinging doors, plan on heavy-duty hinges and hardware so you don't have to keep re-hanging them every few years. While expensive, a full piano hinge works on large swinging doors.

Sliding doors offer space-saving advantages as well as good climate control. They can be left partway open for a bit of light and air without having to be propped awkwardly ajar. With proper installation, they can survive high winds.

Whatever type of door you eventually decide on for anywhere in the barn, be sure that your doors:

☑ Open wide enough to allow you to transport machinery, feedstuffs or other equipment through them

☑ Are wide enough for you and others to safely lead animals through them

☑ Have adequate drainage near them so the areas of high traffic do not become mud puddles during inclement weather. You may have to add some special footing in doorways to reduce mud problems. Rubber mats work well, either bought specially for stalls, or made from old conveyor belting, or purchased from the manufacturer. Concrete or gravel in front of the doors and up to the front of the stalls prevents the soil from wearing down to a muddy trench.

Inside the Stall

Because the inside of the stall is going to take a lot of abuse from your horse, plan to install just a few key fixtures. Protruding objects such as hay racks and storage boxes are an invitation to chewing, and your horse may

get injured bumping into them. Better to have removable stall fixtures, or round the corners on those you cannot do without.

The most natural arrangement for feeding hay is to throw it on the ground. Horses are designed to eat from below their body level, and this way no dust or fragments fall down into their eyes, ears and nose. However, to keep down parasite infestation, you must keep a good clean stall if you feed off the ground. If you have an inveterate stall walker who tramps his hay into the bedding, try using a hay net (hung at the horse's eye

An alternative to rollers for holding the bottom of your sliding door steady is this rail and groove arrangement, which keeps the door from swinging. (Designed by Fritz Brittain of Albemarle Barns, Charlottesville, Virginia). Photo by M. F. Harcourt

cleaning, and some with rims prevent eager horses from spilling the grain onto the floor. Buy one, rather than building one yourself: The wear of the horse eating will soon turn up splinters, the grain worked into the wood encourages wood chewing, and you cannot clean or disinfect a homemade box efficiently.

Your horse needs at least an ounce of salt per day, and a salt brick is the easiest way to satisfy him. Three types of salt racks are commonly available: a cage with plastic-coated wire, all plastic, or a metal plate with edges that catch the side channels in the brick. These are adequate, although once the channels break off the brick from wear, the metal plate model is useless. You can always keep the salt brick in the feed bin, though, no matter what shape it's worn into. Or you can buy the 50-pound model and set it on a board on the floor in the corner. It will last a very long time.

For watering, you can install an automatic waterer in the corner, or simply hang a rubber or plastic bucket from a ring in the wall. Buckets freeze in the winter, and ice is easier to break in a rubber bucket, although they are heavier. Rubber is also less prone to split from thermal shock, such as when you pour hot water in it on a cold day. When hanging the bucket, remember to use snaps or links that have no protruding points on which a horse can catch his face or halter.

Here's a tip the Dutch use for keeping horses watered on frozen nights: After the evening feeding and watering, twist a handful of hay into a short rope and set it in the water bucket so that one end sticks above the water level. Then, in the night as the horse goes to drink from the now-frozen bucket, he can tug the ice free as he fiddles with the "hay handle."

If you plan to groom your horse in the stall, you can manage quite well by

A simple laundry sink can be used for cleaning everything, from tack to Jack Russell Terriers. Photo by M. F. Harcourt

level and removed when empty) or install a round-cornered hay rack in a corner where he's less likely to catch a hip or his head in passing. Place it so the horse doesn't get a face full of dust and hay as he eats. Avoid the temptation to put a rack at floor level. The danger of the horse getting caught in it is too great to risk, even if your horse is the calmest of creatures.

Feed bins offer many choices. Good corner models are available in every price range, most can be removed for putting the lead rope over your horse's withers and letting him move as you groom. If you find yourself spending more time saying, "stand still" than grooming, you might place a tie ring in the center of one wall at the horse's eye level. Be sure to tie your rope to a loop of easily broken baling twine, though, because a panicked horse in a small space like a stall is at his most dangerous—you can't get out of his way.

GRAIN STORAGE

Grain storage should be close enough to your main stall area to be convenient but separated from your horses' reach by a barrier of some kind. Horses are famous for opening feed rooms and storage cans and eating enough grain to kill themselves.

Your storage area doesn't have to be fancy, just secure. Grain can be stored in everything from garbage cans to built-in bulk storage containers lined with metal to keep out rats and mice. Grain storage should serve these purposes:

☑ Easy access for feeding convenience

☑ Protection behind a horse-proof barrier

☑ No access to rodents and insects so they can't spoil grain

☑ Preservation of freshness, free from mold or mildew. Mold and mildew, as well as rodent and insect residue, can cause illness and even death in horses.

One of the most efficient grain storage containers is the plain, old metal garbage can, provided you run a heavy spring, chain or bungee cord through the side handles and over the top to make it horse- and rodent-proof. A 30-

If bulk feeding is your plan, a chute such as this could be ideal. Placed in the loft, the main feed box, made of galvanized iron, pours grain into a cart when the hatch is opened. The cart is then wheeled down the row of stalls and grain is scooped from there, saving the labor of carrying many buckets up and down the barn. Another advantage with this arrangement is that there is never a large supply of grain within a loose horse's reach. (Windchase, Hillsborough, Virginia). Photo by N. W. Ambrosiano

gallon can holds 100 pounds of most grains and can be cleaned, moved and even painted as changes in stable management demand. To lengthen the life of these cans, set them on wooden pallets or a sheet of plywood unless they

In this simple feed and equipment area, metal garbage cans hold grain ready for feeding while bags not yet stored away are safe from dampness on a concrete floor. Tool hooks on the walls store cleaning equipment safely out of the way and give the barn a workmanlike appearance. (Takaro Farm, Middleburg, Virginia). Photo by N. W. Ambrosiano

are on a concrete floor, as moisture beneath them will eventually rust a hole in the bottom.

To provide a more polished look to your feed room, you can build a counter high enough to set the cans under it, and slide each one out as you need it. Or run two boards along the floor beneath the counter on which to set the cans. Make the back board higher than the front one to allow the can to angle slightly outward for easy grain scooping. If you like, you can then set a third board against the front legs of the counter just high enough to keep the cans from tipping over, but low enough that empty cans can be lifted out.

If you can lay hands on an old chest freezer, that makes a very secure grain bin, but you must dispose of the original door lock. Bins full of grain are magnets for children, one of whom might become trapped with tragic results. Replace the door lock with a metal hasp that is installed upside down—with the flap on the bottom—so it must be lifted into place to lock.

Some extra features in the feed room can be worthwhile. The counter, mentioned above, gives you a working surface for mixing grain and supplements, pounding medicine boluses and general work. A light and an electrical outlet are great pluses, too. The outlet allows you to plug in an immersion heater to make hot water for mixing bran mashes or scrubbing buckets.

HAY STORAGE

Hay storage is a vital part of farm planning, whether you have one horse or 100. Even if you choose to buy five bales at a time from the local feed store, you need a good place to keep them while they last. Your storage area need not be rodent-proof like a feed room, but it must offer some protection from water—both rain water and soggy ground. To prevent mold problems from underneath the bales, find some wooden pallets or build a simple grid, strong enough to support your weight, to use as flooring beneath the bales. The decent ventilation provided by this will keep your hay fresher longer.

Remember that hay is a fire hazard as well, so line your storage area with fire-retardant materials that will hold back flames for at least one hour. You can buy ⅝" fire-rated sheetrock from

building supply stores; check with them on other locally available materials.

In planning the hay area, take into account the amount of hay you will buy at a time, both now and as far in the future as you can forecast. Figure on volumes of approximately 40 bales per ton, which take up about 260 cubic feet of storage space.

Buying hay directly from a farm can be most economical if you know how to select quality hay and know a realistic price for your area. Here again, your local Cooperative Extension Service can offer education on the subject as well as a list of suggested hay growers and sellers near you.

Sources of hay from least expensive to most expensive include:

 Hay purchased directly from the grower

 Hay purchased from a broker who buys from farmers in a wide area and sells in bulk locally

 Hay purchased from a local feed store

While feed stores almost always charge a delivery fee, sometimes both growers and brokers will also deliver for a small additional fee. If you pay someone else to unload and store the hay, it will add to the cost. Remember that growers and brokers prefer to deliver in large amounts such as 1 to 10 tons.

If you do your homework, you can find a reliable, cost-effective source of hay that fits your overall operational plan. Check with other horse owners to see if a number of you can buy several tons at one time. Then a tractor-trailer load (10 to 15 tons) can be delivered to a central location with each farm owner responsible for the immediate transport of his or her own share.

Remember, too, that hay is a seasonal commodity. It's cheaper in the summer when haying weather is good than in the dead of winter, so plan your

Here is a properly installed hay rack, with no sharp points to hurt a horse. The hay rack is at a horse's approximate eye height, and low enough that dust and particles will not fall into the horse's face as he eats. (Takaro, Middleburg, Virginia). Photo by N. W. Ambrosiano

These trap doors in the hayloft are centered over the hay racks in the stalls below, allowing hay to be shoved easily through to two horses in one effort. The doors are drawn open with heavy string that runs from the door through a ring in the rafter and tied at the end with a large washer, so you never need scratch around the loft floor looking for the door handle. In addition, the paths to each trap door in this barn are marked on the loft floor in red paint, so as new loads of hay are brought in, they can be stacked out of the pathways. (Windchase, Hillsborough, Virginia). Photo by N. W. Ambrosiano

storage accordingly. If you use a local source, make sure he or she has enough to sell you throughout the winter or will reserve hay for you to purchase during the winter months.

Let's add a word about management here. Through education, your Cooperative Extension Service can help you decide the type and amount of hay you need. Not every horse must have the most expensive hay. Knowledge on the subject will keep you from spending money unnecessarily and help you maintain your horse in good, healthy condition.

As mentioned earlier, hay should not be stored above your animals' stalls if at all possible. Overhead storage is a fire and health hazard that can be eliminated by simply having a separate storage area specifically for hay (and bedding or other flammable necessities). Studies have shown that dust particles play a large role in the upper-respiratory infections often seen in today's stalled animals. No matter how high the quality of the hay you purchase, there will be some fine dust and particles that will add to the other airborne particles present in

your facility. Also, on a practical note, anyone running a barn soon discovers that retrieving hay from a 10' high loft is no fun, even if you have a hay conveyor available.

Nevertheless, if you plan to have overhead storage, take the following precautions:

☑ Fit the boards or plywood sheets in the hayloft floor very tightly together to reduce the amount of dust and hay particles that rain down on your animals.

☑ Do not make the loft floor and stall ceiling two separate structures in the hope that this will eliminate the fall of residue. All you are doing is creating the ideal spot for a fire to start as dust and debris collect between the floor and ceiling.

☑ Consider adding fire-rated sheetrock under the flooring.

Round Bales

You've seen elephant-sized round bales of hay being fed to cattle or spread like huge grassy pillboxes across hay fields. While round bales appear unwieldy, it can make good sense to feed hay this way depending on your setup. You'll find you spend less time offering flakes by hand. Round bales can be very economical, saving both in purchase price and labor charges. But there's a trick to using these massive bales in a horse operation.

Round bales are just what the name suggests: cured hay rolled up like a jelly-roll cake. They range from 800 to 1,200 pounds, but one person with a tractor and the right gadgets attached can bale, move and store them with ease. From storage, they can be placed in a feeding station for free-choice dining for your horses. This can be ideal if you have an erratic schedule and no help, if you're trying to keep a steady intake of food available through the day without good pasture, or you're just unwilling to waste time hauling hay when you could be riding.

Choosing Your Bales

In choosing round bales, you do have to be more cautious about the quality of hay provided, as most round bales are produced for cattle that can get by on lower-quality forage than can horses. Be sure to ask specifically for the same types of hay you'd order in square bales, and check that it is weed free and cured properly.

Curing is an important issue in keeping round bales once the hay's rolled up. Correctly cured bales help ensure your horses aren't put off by mold or don't develop digestive and respiratory troubles from moldy or decaying hay.

Feeding Efficiently

To feed a round bale efficiently, you need to place it in a round feeder or rack (commercially available at larger feed stores) to avoid waste and trampling. Use a pallet or tires underneath the

bale to keep it from wicking moisture up from the ground, and place the feeder on high ground if possible to avoid a swampy, trampled area as the horses feed through the season. Also shift it from one area of the pasture to another to avoid excessive damage to the land around it.

For best maintenance of hay quality through each bale's lifespan, cover the top with a tarpaulin securely strapped down, and then as the bale shrinks with use, set a board or plywood across the top of the rack to support the tarpaulin. This keeps water from pooling in the tarp, potentially leaking into the center of your bale.

Round bales can be easily incorporated into your overall management plan. They are not only economical but reduce labor while providing horses a constant source of roughage. (Montgomery Property, Ashley Heights, North Carolina). Photo by M. F. Harcourt

Hay, Bedding And Feed Storage Space Requirements

Material	Weight Per Cubic Foot in Pounds	Cubic Feet Per Ton
Hay - Loose in shallow mows	4.0	512
Hay - Loose in deep mows	4.5	444
Hay - Baled loose	6	333
Hay - Baled tight	12	167
Hay - Chopped long cut	8	250
Hay - Chopped short cut	12	167
Straw - Loose	2-3	1000-667
Straw - Baled	4-6	500-333
Silage - Corn	35	57
Silage - Grass	40	50
Barley - 48# 1 bu.	39	51
Corn, ear - 70# 1 bu.	28	72
Corn, shelled - 56#1 bu.	45	44
Corn, cracked or corn meal - 50# 1 bu.	40	50
Corn-and-cob meal - 45# 1 bu.	36	56
Oats - 32# 1 bu.	26	77
Oats, ground - 22# 1 bu.	18	111
Oats, middlings - 48# 1 bu.	39	51
Rye - 56# 1 bu.	45	44
Wheat - 60# 1 bu.	48	42
Soybeans - 62# 1 bu.	50	40
Any small grain*	Use 4/5 of wt. of 1 bu.	
Most concentrates	45	44

(Courtesy: American Plywood Association.)

* To determine space required for any small grain use wheat (60# = 1 bu.) for example. Then : 60 (4/5) = 48# wheat per cubic foot volume. to find number cubic feet wheat per ton. Then:

$$\frac{2000\# \text{ (Wt. of one ton)}}{48\# \text{ wheat per cubic foot volume}} = 42 \text{ cu. ft}$$

The lifespan of your bales will depend on the hay quality, other forage and grain available and your horses' appetites. An example, however, might be that of one author's stock: three mature pleasure geldings, fed two grain meals daily, who polish off a 1,000-pound alfalfa bale in two weeks.

Commercial round bale feeders come in several varieties, some better for horses than others. One that works well is in three metal-pipe sections that bolt together, and has a full panel surrounding the lower portion of the rack. This keeps stray horse legs out of the frame and yet doesn't add excessively to the weight. Such a feeder weighs about 120 pounds, light enough to be shifted by one person without taking it apart into sections.

The racks are likely to shift somewhat during use, as the horses bump them (less so than cattle, however) and if the feeder openings are too low, the horses will rub their manes as they feed. You can buy one with legs, adjust the height of the entire feeder by raising it on tires, or choose a "tombstone" feeder whose feed openings curve upward like the profile of an old tombstone. This allows horses to raise their heads and chew comfortably without bumps or rubs. If you choose a model with legs, you have the advantage of a longer-lasting feeder, since there is less metal in contact with damp ground to rust out.

Some management differences should be taken into consideration when using round-baled hay, as the outer few inches may be weathered or, if the hay was bagged, even slick or moldy. Be sure to strip off the outer layer initially, and check the bale as it progresses. The horses will naturally choose the more appetizing inner portions to eat, dropping second-quality sections to the ground. If this discarded material is merely slightly weathered, it does no harm and can be eaten later. If it's moldy, however, and the better hay is eaten away with no replacement, your horses may resort to taking the moldy stuff and risking their health. Simply providing a huge hay pile does not remove your responsibility as horse owner to check your stock daily and examine their food.

An efficient tack room with good use of wall space. (Takaro, Middleburg, Virginia). Photo by N. W. Ambrosiano

Storing Round Bales

As with regular bales, round bales are strongly affected by the way they're stored before use. Left uncovered and on the ground, they can lose about 63 percent of their edible portion if the weather is wet and warm. Stored in a barn, less than 14 percent is likely to be wasted. Calculating the cost of lost hay to the cost of a storage building over time, you can save about $1 per bale by storing it under cover. An added storage building on your property is useful to have anyway, whether it's filled with hay, horse trailers or equipment.

Minimizing Losses

Storing your bales covered, then feeding them covered as well, will give you the most palatable hay in the long run. Try changing hay types through the seasons as well, though, to keep your horses interested. (Remember that any drastic changes in hay, such as grass to legumes, must be done gradually to minimize risk of colic.)

While feeding good quality hay through the winter is sensible, save a little of that good stuff for early spring, to reduce pressure on pastures that are just beginning to grow tasty new shoots. If you're offering old, low-quality hay, the horses will reject it in favor of the new shoots, and you'll waste hay while incurring damage to the pasture at a fragile stage. Tempt your stock back to the feeder with good alfalfa or top-quality local hay. Even if it requires a temporary restraining fence, let your pasture get a good start before you turn the grazing machines loose on it.

TACK STORAGE

Your options for equipment storage are almost unlimited as you plan your barn. If you choose to devote all your space to horses and not horse gear, you can build all stalls and reserve a spot in the aisle for a storage trunk. On the other hand, you might choose to have a tack room that doubles as a trophy display area and hide the dirty gear in a special cleaning room

Remodeling an old barn doesn't mean you can't have a comfortable and functional tack room. By adding hot and cold water, electricity and a concrete floor to one stall, Linda Daniel created a clean, workman-like environment that opens to an adjacent grooming stall as well as the main aisle. (Brandon, Spring Grove, Virginia). Photo by M. F. Harcourt

For an almost self-sufficient tack arrangement, this heavy-duty sewing machine is the final touch. Set up with plenty of work space, it means worn blankets and gear can be repaired right away. (Windchase, Hillsborough, Virginia). Photo by N. W. Ambrosiano

reserved for scrubbing after every ride.

If everything, such as brushes, tack and so on, has a place, it's easier to keep every thing in its place. With a more efficient operation, you spend more time riding than you do looking for hoof picks, for instance.

For the average horse owner, one room will suffice, with wall-mounted saddle and bridle racks for clean leather and a corner devoted to soiled saddlery.

A tack room the size of a stall is fine if you can plan it; it will hold as many saddles as you can put racks on the walls. If you'll be the only one working in it, you'll have room for a work table and more.

Key ingredients for a workable tack room are off the-floor storage for your saddles and bridles plus a spot for brushes and another for first-aid equipment. If the rising trend in tack thievery bothers you, make sure you can lock

Saddle racks outside each stall can be handsome metal racks that fold out of the way when not in use. Photos by N. W. Ambrosiano

your tack area securely, and remember to do so. The rest is gravy. You can add many more wonderful things when designing your own facility:

- ✔ Simple hot and cold water taps, a double-sided sink, a washer and dryer for tack and saddle-pad cleaning, and an electrical outlet which allows for a coffee maker, and even a microwave for quick meals on the run

- ✔ A saddle rack that flips up to hold a saddle upside down so you can clean the underside

- ✔ A wide counter, where you can take a bridle apart and lay it out for cleaning, oiling and repair

- ✔ A heavy-duty sewing machine so you can renovate worn blankets on the spot

- ✔ A bandage-rolling spool for winding up freshly washed leg wraps

- ✔ Two bridle-cleaning hooks, one for dirty bridles and one for freshly oiled ones to dry

- ✔ A non-flammable heating system for winter days

- ✔ An air conditioner (no more damp, moldy tack)

- ✔ A refrigerator for keeping medicines, ice packs and the odd chilled soda

- ✔ A light bulb-powered bit warmer for cold days

Any horseperson can give you a list as long as your arm of ideas, things that make good horsemanship not just a matter of conscience and safety, but of comfort as well. When you tour people's barns, ask to see their tack areas, and find out what might work for you.

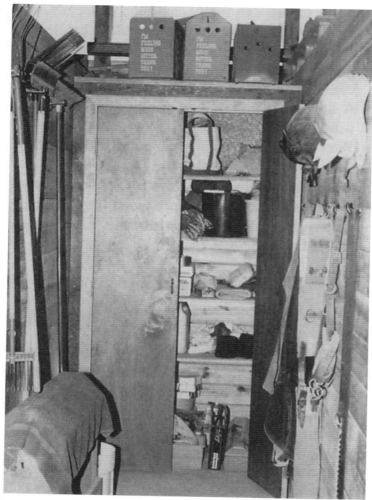

If you split a stall space into several storage areas, let one of them be an open aisle like this, with a cabinet at the end. Lined with cedar-chip board and made with widely spaced shelves, this cabinet was designed for moth-proof blanket storage, but holds much more. Plenty of hooks on the walls ensure a tidy place for all sorts of equipment, where it is out of the barn aisle but still accessible. Photos by N. W. Ambrosiano

GROOMING/ WASH STALLS

A wash stall is a bit of a luxury for the casual horse owner, but for anyone involved in hard training, showing or breeding, it's a necessity. The term "wash stall" is actually a misnomer. What is really needed is a spot with good drainage, hot and cold water, and crossties. That can include aisleways and corners of the barnyard, if simplicity is to your liking.

Drainage is the first requirement, as it takes a lot of water to wash a horse. Once it comes off the animal the water must be drained out of the traffic areas. If you choose to install a subfloor drain in a concrete pad, be aware that clogging will be your main nuisance. Make sure the grating over the drain is sturdy, removable, and is cleaned frequently, and use a drain with a sediment bucket. Also look for a fiberglass drain grating.

Another method is to slope your footing to one wall, and run a 3" pipe from the lowest point out through the wall. On the outside, don't hope for nature to handle the runoff, or you'll flood your barn's footing. Either extend the pipe away from the barn, or drain it into a drywell filled with gravel.

One northern Virginia eventing stable went with the concrete floor option, but added a foot-soaking area for hard-working horses. Along one wall is a 6' x 4' soaking pit approximately 4" to 6" deep. The rest of the stall drains into this pit, and the pit drains into a wall pipe as described above, but there is a plug to block the wall pipe during treatments. Horses who won't tolerate a foot in a bucket will stand happily for warm- or cold-water treatments in this open bath.

Concrete is not essential for a floor; gravel over a good base will do. Or, if you wash in the barnyard, a good thick patch of grass that drains well makes a fine footing. Just be sure you won't be standing in a puddle or a mud hole by the end of the bath.

You can do without hot water, especially if you're in a warm climate, but it makes frequent bathing and other barn chores more pleasant when the mud is thick on the horse and a chill is in the air.

You'll be happiest if you make your tie space a crosstie or next to a wall so that your horse can't roam around the tie ring during the bath. Opinions differ over where to place tie rings: at wither height or above. In any case, the rings should not be any lower than your

horse's withers. Otherwise, he could pull back and damage his neck if tied too low.

The ropes or straps you choose for tying can be anything that allows a quick release in case of emergency. Some people prefer chains, but a horse throwing his head can toss the chain into your face, which is sure to make you a devotee of soft cotton ties. Loose ties are a liability, so measure them carefully. They are about right if they just reach to the horse's halter from either side. Any looser and the advantage of cross-tying in the first place is lost. The horse will be able to turn around or even get a front leg over low ties.

Two schools of thought govern the attachment of ties to the wall. Some prefer strong ties bolted firmly to the wall, with a quick-release catch at the halter. A panicked horse can be released if you deem it necessary and you can get to his head. Yet he can't idly pull himself free.

This tack-cleaning area has it all: a washer and dryer for wet pads and wraps, cupboards for storage of cleaning gear, and plenty of hooks and racks for the saddlery. (Takaro Farm, Middleburg, Virginia). Photo by N. W. Ambrosiano

While aisle ways should be clear and horse proof, some clever storage allows access to essentials at all times. In this barn, a horse in the grooming/wash stall need not be left alone, as everything is nearby. Exercise wraps and cottons are in the open racks on the walls, saddles wait on the folding racks at center, blankets are folded on racks along each sliding door, and the hanging mesh basket near the horse's head holds sponges and brushes where they can dry between uses. (4th Estate, Paper Chase Farms, Middleburg, Virginia. Designed by Upperville Barns). Photo by M. F. Harcourt

This concrete-floored wash stall has everything close at hand, yet out of a horse's reach. The diagonal shelves at the rear of the stall have no protruding corners to hurt a horse, and the thick cotton cross-ties ensure that he stays at the front of the wash area. Drainage could be slightly improved, though. Notice that the water pooling in the corner has discolored the walls. Pressure-treated plywood is a must, even where the walls do not touch the outside soil. Photo by N. W. Ambrosiano

Others choose to link the ties to wall rings with a loop of baling twine, letting the horse break the twine before hurting himself. Your management style and the temperament of your horses will define your best course here. If you're not sure and you'd like to avoid replacing halters, go with the baling twine option.

A newer product has come on the market that also addresses the needs of halter-breaking horses. Known generally as bungee ties, these can be found in lead rope, trailer-tie and cross-tie lengths and they combine a thick elastic core with hook-and-loop fastener attachments and extremely strong carabiner snaps on the ends. Unless your horse is seriously pig-headed and sits back to the end of the barn, he's likely to calm down when the rope has some give to it, and he'll stop fighting before the hardware gives out.

If you plan an enclosed wash pit, a fairly standard size is 10' x 10'. This gives you enough room to work around your horse without him crushing you against the wall. If you plan a barn with a stall or

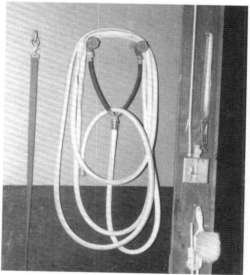

These wash stall details are well-arranged here. At left, a nylon cross-tie is attached with a quick-release clip to a screw eye in the wall. Beside it, a hose is safely coiled over hot-and-cold water taps, which are out of reach of a horse's legs. To the right, the essential scraper, hoof picks and brushes hang neatly off the floor, along with an emergency switch. The wall switch for the overhead light is safely enclosed in an exterior-type housing, and the covered plug allows the area to be a clipping stall once the water is drained away. Photo by N. W. Ambrosiano

Good use of a "dead" sloped area. Owner Bob Perks put in two wash pits for his boarders. (LaGrange Farm, Bob Perks, King George, Virginia). Photo by M. F. Harcourt

Need a covered wash area, but can't find the space? Linda Daniel turned part of her aisleway into a wash area by simply digging out some dirt, bracing the area with 2" x 4" (or 4" X 4") and filling in with large, round gravel. She doesn't lose access to her run-in shed or pasture and yet has a convenient, usable wash area with little to no mess. (Brandon, Spring Grove, Virginia). Photo by M.F. Harcourt

two more than you have resident horses, you've got the beginnings of a wash pit right there. If you're sharing space with a storage area or enclosing the end of the aisle, just be sure to give yourself enough room for a full-size horse and you. Some horses are claustrophobic in a small space, and you need to have enough room to maneuver in case of trouble. If a narrow space is all you have available, make sure you can at least turn a horse around in it without scraping hips and sides.

Since giving a bath is more than merely hosing down the horse, equipment storage is a big help. You need plenty of space for wet sponges and brushes, a place other than underfoot for the scraper, and a reel or rack for the hose. Hanging metal mesh baskets make good gear keepers, as do plastic milk crates. Wooden boxes tend to become waterlogged and rot, and things in the bottom can mold instead of drying. With a little scrap lumber, you can build a set of open corner shelves in the back corner of your wash stall deep enough to hold a few items and reinforced with a piece of molding across the front to prevent the contents from rolling out.

Keep your hose rolled securely and out from under the horse's feet to ensure both its longevity and yours; a tangled horse or groom is counterproductive. Just be sure that if you install your hose reel or hanging rack along a wall, your horse cannot hit himself on protruding parts.

Good light is essential in the wash area, and important that it be out of reach of a rearing horse. Bathing and associated care are more likely to bring on violent reactions than most other kinds of handling.

HIGH-TECH AND SAFETY FEATURES

There are many new high-tech and safety innovations that you can add to your barn if you desire.

Fire Prevention

The worst nightmare of every barn owner is fire. It starts suddenly, drives normally staid horses into a suicidal panic, and sweeps a barn before you can begin to stop it. To keep it as unlikely as possible, you must build in safety measures and practice good management.

Sprinkler systems can be easily incorporated into the overall barn construction. This option will pay for itself in insurance savings if you have a public-access stable. Dry-pipe sprinklers drain back when not in use, a very helpful item in cold climates.

If you have living quarters adjacent to or over your barn, you might want to consider some of the newer types of alarms to warn inhabitants of potential dangers before it's too late to evacuate or they become unconscious from carbon monoxide poison or smoke inhalation.

In addition to conventional smoke alarms, you'll find alarm systems that activate when there is an abnormal rise in the air temperature. These can be effective as they sense heat increases often before smoke is sufficient to set off the alarm.

Carbon monoxide sensors are available to indicate a high level of this dangerous gas. These can be useful for not only fire detection but for faulty heating systems, or when it's necessary to run gasoline powered equipment in an enclosed area. Carbon monoxide poison sneaks up on man and beast, putting them to sleep before killing.

Building with fire-resistant materials or specially treated lumber can also add to your peace of mind. Find out what is available from your local building supply store, and check with your fire department for recommendations. In most areas, fire barriers as walls next to living quarters need to hold back flames for two hours. For barn ceilings, the requirement is three to four hours if an upper living area exists. Combining fire-retardant lumber with ⅝" fire code sheetrock (each sheet of which resists burning through for approximately one hour) can meet this requirement.

It's best to plan for the worst case scenario when deciding how much protection to build for, as you want to have adequate time to evacuate yourself, your family and your horses in the event of a fire.

It's best to work closely with your local building code inspectors to plan the amount of protection required by your local building codes before you build. You'll find variations in requirements, and codes are often open to interpretation. Inspectors can help you protect yourself; design a plan that meets their guidelines and your own personal protection requirements if living areas adjacent to or over your barn are part of your plan.

Make sure your doorways are easily opened and accessible in case you need to evacuate horses quickly. All the latches should be operable with one hand. If you plan on keeping more than 10 horses in the barn at once, be sure at least two clear exits are available at all times and that Dutch doors can be opened from the inside and the outside.

Don't skimp on fire extinguishers. You don't need the big, unwieldy institutional ones, especially if you expect small people to use them. Just make sure the ones you get are fully charged and in places that everyone in the barn knows about and can reach. Check with local officials on number and placement of extinguishers.

No matter how good a housekeeper you are, take another look around the barn. Get rid of hanging, dusty cobwebs that could flame up and loose piles of hay or bedding against the walls. For the safest arrangement, store bedding in a separate shed. If you have floor-level storage or even a loft, consider lining the floor, walls and ceiling with fire-retardant materials that will hold a fire back for at least an hour.

While a hayloft is not generally recommended, one advantage of it is that fire burns upward. Hopefully, that will give you time to get the horses out before beams begin falling into your path.

A hose with a good sprayer head located at the midpoint of the barn can be a lifesaver. Even if you have automatic waterers and a wash stall at one end, go ahead and add this extra faucet to the plans. Put in as much fire protection as you can afford; it will be worth the cost in peace of mind.

Don't minimize the importance of a phone in the barn. Calling the fire department is your first step in event of a fire. No matter what, you'll need to give them a jump on the flames.

Security

You'll find numerous security systems that will allow you to sleep peacefully at night, safe in the knowledge that your horse has not been taken for ransom or sold to the nearest meat market. In addition to horse theft, stealing tack and related equipment is big business because such materials are so hard to trace and so easy to transport.

The cheapest security option is a dog with a good bark or, even better, geese or guinea hens. If you haven't been exposed to these latter two organic alarm systems, be assured they can make enough noise to raise the dead. If you have truly valuable animals or a facility where you're responsible for horses and equipment belonging to others, an electronic security system may be necessary.

You may want to add an electronic security system for several reasons, and it's best to plan accordingly during the early building stages. Check with a reputable company to see if they can devise a system for your barn. Many home systems involve motion detectors, which pose a possible problem if you have dogs, cats, birds or horses milling around at night, so be sure the company understands this. Then, see if you need to plan for any additional electric lines or back-up batteries to accommodate the system being installed.

Intercoms

Intercoms can be part of your security system and can also let you monitor ongoing activities in the stable. For example, they are quite handy for checking on mares about to foal or colicky horses or for catching the horse that has a tendency to cast itself. Of course, an intercom also makes it easier for you to keep track of the kids, call folks for dinner, and more.

Intercoms are very simple to purchase and install yourself, or you can have them installed for you. Your pocketbook and needs will determine your choice. For little expensive, you can install a system usually sold as a baby monitor that will allow you to keep an ear in the barn, even though it's only a one-way device. Two-way systems are easily available as well, and you can even install a complete loudspeaker system that allows you to use your facility for horse shows. You can find intercoms at consumer electronic stores, or check catalogs for special packages.

A Phone in the Barn

A phone in the barn is an absolute necessity these days, even if it seems a bit frivolous when you're there to ride, not talk. It can be a traditional phone or a cordless model that can go into the arena with you. If you use your barn as a refuge from the phone, you can always turn the bell off. After all, you really need the phone for outgoing emergency calls, especially if you don't get cell service from your barn. Just in case of fire, horse emergencies or just a need for general communications, hook one up.

Closed-Circuit Television

Determining if a broodmare is about to foal just by listening can be tricky, but seeing what a mare is doing on a closed-circuit television leaves no doubt. Here again, this piece of technology has its place in certain facilities, and may be incorporated as part of the security system.

Closed-circuit television systems are expensive if you have a small-scale operation. But today's technological advances give you a wide range of options and price ranges. If you have good mares bred to expensive stallions, you might consider this as an option to hanging out in the barn all night waiting for a sneaky mare to foal.

Construction

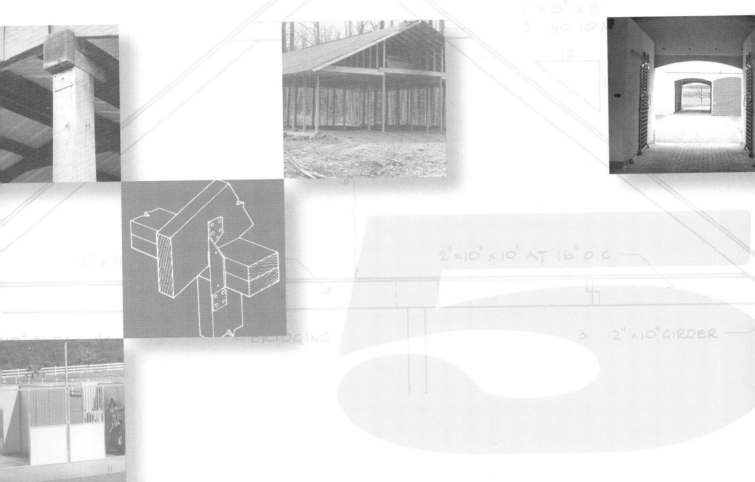

Obviously, if you plan on building your facility yourself, you must have good construction skills and knowledge. While you can learn some things as you go, it's not a good idea to start with zero knowledge on construction techniques.

If your facility will be constructed by someone else, some knowledge on construction methods and concepts can help you understand what's going on and ask good questions. And since you are the one paying the bill, you have the be right to be satisfied with what is being built for you.

The early stages of your planning should have included finding an architect, builder or subcontractor you can work and communicate with. If you are the main builder but plan to hire out certain parts of the project, you will need several subcontractors to work on your facility. Keep in mind that, most likely, the person who knows the most about fence installation is not your best choice for digging your wells.

While the following section won't make you a master builder, it can give you enough basic knowledge to allow form and function to come together to give you a perfect place for your horses.

STEP-BY-STEP PLANNING

You have selected your land and now need a step-by-step list of things to do to head you in the right direction. The following procedures will get you going on the right track.

STEP 1: Check out the zoning regulations, municipal building codes, covenant restrictions and any other stipulations that may affect your plans. Your area may have regulations concerning land use and water sheds, and pollution from soil runoff that you need to investigate. Ignoring the details of these laws can prove costly later on or can stop your construction dead in its tracks.

Suggested sources for this information are:

- Municipal zoning/planning offices
- Building inspector's office
- Soil conservation service

STEP 2: Make appointments at the appropriate information offices, and es-

tablish a positive relationship with the people who work there. A run in with an inspector can sour your whole project and delay every step. Ask the inspection staff for their advice on your project, and follow it. While they may not be horsepeople, they know the local building scene and can steer you away from potential problems.

Set up a file system so that you have easy access to the information you gather. Departments and service offices available to you are:

- Cooperative Extension Service
- Well-respected local horse owners
- Equine consultants
- Natural Resource Conservation Service
- Forestry Service (if your land is wooded and you need information on getting the timber removed)
- Agricultural Stabilization and Conservation Service
- Builders
- Architects

STEP 3: Network with the local horse community. Find out about groups, organizations and educational meetings at local feed and tack shops. Get on their mailing lists.

STEP 4: Subscribe to one or two quality horse magazines that give you current information on horses and horse management. (As your experience level increases or your style of riding or direction in the industry changes, you may want to drop some and subscribe to others.)

STEP 5: Begin asking yourself what you want, need and must have for your operation. If your budget is limited, make a list of what's essential to start your operation and what can be added later.

STEP 6: Locate a barn builder in your area who has a good reputation for quality work and reliability. Look at several barns he or she has built and talk to the owners. Never hesitate to ask for references or referrals from your builder. While many well-qualified builders may work in your area, we strongly recommend someone who has built horse barns and/or is a horseperson too.

STEP 7: If you plan on being your own builder, locate subcontractors. The exception to this would be if you are confident about your hiring, concrete-laying and woodworking abilities. Building the facility, if it is on a small scale, is not difficult. However, if you are not 100 percent sure about the electrical and plumbing aspects of your barn, consider subcontracting.

If you plan on being your own builder, you may find it worthwhile to pay a local barn builder for some of his or her knowledge, using the builder on a consultant basis. The builder's knowledge of local building codes, soil problems and material availability will prove invaluable.

STEP 8: Locate and price local firms that sell lumber and other building supplies. Here again, begin networking. Establish an account and credit with a good company so you can plan a budget for your building schedule. The store can also be a good source of how-to information as well as a source of suggestions for materials and substitute items you will need.

STEP 9: Invite any local Natural Resource Conservation Service or Cooperative Extension personnel willing to make an on-site visit to come and make recommendations as to location of facilities, angle of placement in relation to prevailing winds, suitability of soil and

so on. Then lay out your plans on paper to get an idea of the movement of horses, machinery and pedestrians.

STEP 10: Think about drainage before you put down the foundation. If you're going to build up your barn site, investigate local materials that will be economical to provide the dry flooring you need in the barn. Will your current fill material be suitable, or will you want to add drain lines in the stalls? Plan to dig these before adding the fill. Put in any drain lines you need for the wash pit, washer/dryer and bathroom if you plan on adding them either now or later. You don't have to put these appliances in right now, but if you haven't made arrangements for them, it will be both more expensive and more trouble to do so later.

STEP 11: Begin construction using the building order list that follows. Monitor your progress daily and be ready to change plans, if necessity demands. Have a scheduled completion date, and if you're using a builder, work with this professional to meet the building deadline. If you are building the facility yourself, the builder will help you stay on schedule.

Building Order List

Once you have selected the construction site for your ideal barn, follow these steps to aid you in completing your construction in a timely and logical order:

STEP 1: Base your site selection on:

 Drainage

 Prevailing winds

 Accessibility to roads, rails, airports or other forms of transportation necessary to your operation

 Availability of water, utilities

- ☑ Passive or active solar considerations
- ☑ Future improvements, additions
  - ■ Arenas, riding trails
  - ■ Parking
  - ■ Fencing
  - ■ Ponds
  - ■ Facility expansion

STEP 2: Prepare your site by:

- ☑ Excavating, grading for roadways, and drainage
- ☑ Determining preliminary utilities layout
- ☑ Acquiring building permits
- ☑ Installing construction utilities (electric, water)

STEP 3: Construct your facility:

- ☑ For pole barns
  - ■ Mark exterior building lines, set batter boards, excavation stakes
  - ■ Dig drainage trenches
  - ■ Mark pole locations, dig holes
  - ■ Set poles, pour footings, if required
  - ■ Backfill holes, tamp in carefully 6" at a time
  - ■ Attach and level splash board, girts
  - ■ Add additional fill for stall drainage
  - ■ Begin roof work
- ☑ Set eave girts, or, if resting a top plate on leveled pole tops, attach it
- ☑ Set rafter girt, if used
- ☑ Step off rafters for gable roof and cut lines where rafters will meet plate or eave girts. If using prefabricated trusses, strap them in place with temporary framing
- ☑ Attach collar ties to rafters

- ☑ Chalk line along tails of rafters; cut even for fascia board attachment
- ☑ Frame eaves with fascia boards and soffits, leaving space for vents
- ☑ Deck and cover roof with chosen material
  - ■ If planning a hayloft, lay joists on eave girts or suspend them from girts with metal joist hangers
  - ■ Lay plywood loft floor
  - ■ Place flooring in tack rooms, feed areas
  - ■ Nail interior barn walls (rough-cut 1" x 6" or 2" x 6" boards) in stalls, frame for windows, paneling, water lines, and so on in tack room
  - ■ Install windows, doors
  - ■ Add exterior siding
  - ■ Stain, add gutters, finishing touches
- ☑ Conventional foundation structure (This assumes you are building a full house.
- ☑ Delete any steps not applicable to your barn. Knowledge of local inspection requirements is necessary.)
  - ■ Set foundation: batten boards, excavation stakes, etc.
  - ■ Dig footers
- ☑ Form footers, and reinforce rebar, add drainage aggregate
- ☑ Pour footers, remove forms
  - ■ Alternate 1: poured walls
    - ■ Set wall forms, plumb and level, add reinforcement; rebar, add drainage aggregate
    - ■ Pour walls and remove forms
  - ■ Alternate 2: block walls
    Build block walls up from footing to finish floor level; fill all cells in block with grout
    - ■ Install top plate to anchors set in foundation wall

- Back fill foundation
- Construct first floor joist
- Build subfloor
- Any time after block walls are built, French drains may be installed and foundation back-filled
- Erect exterior/interior wall studs
- Install second floor joist, if applicable
- Lay second floor subfloor, if applicable
- Add top plate
- Install prefabricated trusses or rafter system
- Erect roof sheathing

—Construct roof felt and roof

—Install exterior wall sheathing, insulation board, vapor barriers etc.

—Rough-in plumbing, electrical work

At some time during the previous steps, concurrent provisions should be made for drilling wells or obtaining other sources of water; electrical services should be connected to a service entrance; and a septic system should be installed or provisions made for connection to existing sewage disposal system

- Begin exterior siding, masonry
- Add drywall installation

Laying Out Your Barn

Decide which way you want your barn to face, based on your land and prevailing winds. Then note the exact length of the outer edge of each wall by placing a peg where one wall will begin, measure to the other end of the wall, and drive another peg. Run a string from a nail in the center of the top of each peg to mark off a straight line along the wall.

To line up the other walls, use the 3/4/5 rule from your old geometry days: To form a right triangle, use those dimensions, or multiples thereof, and the right angle will be in the corner opposite the 5' side. To be sure of your measurements, multiply this out to at least 6- to 10', and be sure to measure all corners this way. If your first markings are square, it will make life much easier and save you money on extra lumber when you get to the roof.

Once you have square corners, erecting batter boards allows you to remove your string lines to set poles, then replace them without going off-track along the way. In each corner, just outside the outer framing line of the barn, sink three pegs conforming to the corner. Tack two boards along the pegs, and put a nail along the top of each board to attach the string exactly along the foundation lines. Replacing the string accurately after each removal will be no problem with the batter boards in place.

TYPES OF BARN STRUCTURES

Pole Barns

The classic pole barn structure achieves its overall structural stability from poles, usually timber, embedded deep into the earth, similar to a flagpole. Generally, the posts need to be embedded at least 5', in order to develop adequate fixity. They must be continuous up to the plane of the roof, without any splices or hinges. The required size and depth of the poles is determined by the height of the barn, the spacing of the posts, the roof load and wind load acting on the barn, the strength of the poles, and the firmness of the soil at the site. The bases of the poles should be treated to prevent rot. Frequently, the gravity (downward) load on the pole requires that a concrete pad be placed at the bottom of the post holes, to spread

Metal connector plates like these at the top of the posts that join two perpendicular 6" x 6"s can add to the building's expense, but save time and trouble in construction. Photo by N. W Ambrosiano.

out the load on the soil. To fill in the space around the poles, granular material should be used, free of clays. This can be sand, gravel or graded, clean fill. The material should be packed down or vibrated, in 6" lifts.

The most critical section of fill around the poles is at grade level, where the lateral stresses are highest. Therefore, the granular fill should continue to grade level or just above, not stopped short to allow for topsoil. In colder climates, where frost can be expected to permeate, a concrete "apron" around the tops of the poles is unadvisable. Ice lenses can form below, and frost heave can push up on the solid projection or rough edge, creating the possibility of severe damage.

The roof support beams, which support the roof rafters, can then be attached to the tops of the poles. Horizontal wall beams, or girts, can span horizontally between the poles to receive the wall sheathing.

Modifications for Pole Barns

One drawback of a true pole barn is that the size of the poles can get quite large, since all of the stiffness comes from the cantilever action of the poles out of the ground. Providing short diagonal braces, or knee braces, at the tops of the poles, can significantly reduce the pole size needed for the same spacing. These braces connect to the roof support beams. They allow the lateral bracing needs of the building to be shared between the ground level and the roof level of the poles.

Another popular modification to pole barn construction is the inclusion of support poles out of a "sandwich" of 2" x 6" treated lumber. Using three 2" x 6" boards, the builder can purchase 12' boards and laminate them to create a 6" x 6" support pole that is 12' in height. To produce supports higher than 12', additional 2" x 6" boards can be cut to any length and added on to create the desired height. By staggering the cuts in

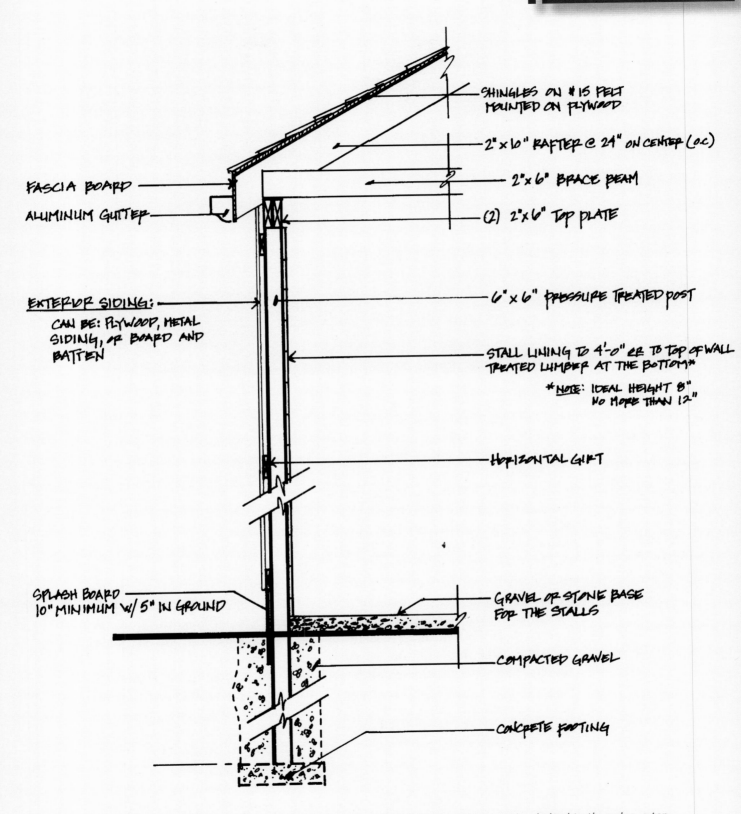

SHINGLES ON #15 FELT MOUNTED ON PLYWOOD

2" × 10" RAFTER @ 24" ON CENTER (OC)

2" × 6" BRACE BEAM

(2) 2" × 6" TOP PLATE

FASCIA BOARD

ALUMINUM GUTTER

6" × 6" PRESSURE TREATED POST

EXTERIOR SIDING:
CAN BE: PLYWOOD, METAL SIDING, OR BOARD AND BATTEN

STALL LINING TO 4'-0" OR TO TOP OF WALL TREATED LUMBER AT THE BOTTOM*

*NOTE: IDEAL HEIGHT 8" NO MORE THAN 12"

HORIZONTAL GIRT

SPLASH BOARD
10" MINIMUM W/ 5" IN GROUND

GRAVEL OR STONE BASE FOR THE STALLS

COMPACTED GRAVEL

CONCRETE FOOTING

This is a section of a barn built with a modified pole construction. Rather than eave girts bolted to the poles, a top plate is used on which the rafters rest.

Joist hangers are the most secure way to handle such large sections of joists and beams where simple toenailing of perpendicular boards would not be enough. Photo by N. W. Ambrosiano.

the laminated pole, structural integrity is assured, and desired height obtained.

Braced Post and Beam Barns

These barn structures are essentially "pinned" at their bases. That is, the barn posts are simply sitting on foundations, and although they are anchored to them, do not have rigid bases like those of pole barns. This leads to simpler foundations, which can simply be sized for gravity load. Their depth needs to reach only to inorganic virgin material, or below maximum frost depth, whichever is deeper.

As the name implies, a braced post and beam barn needs above-grade bracing elements to resist sway. This usually takes the form of diagonal members between the posts. Frequently, roof trusses provide bracing at the tops of the posts.

Stick Construction

This refers to normal house-type construction, where the "sticks" are vertical studs in bearing walls. With sufficient sheathing such as plywood rigor-ously attached to the studs, the walls act as a diaphragm that can provide lateral stability in the plane of the wall. Under each of these walls, a foundation wall is required, with a strip footing below.

How a Pole Building Comes Together

A pole building is so called because it is built around poles or posts that are set in the ground. Rather than sitting on a foundation, the barn in effect hangs from a frame of pressure-treated poles or lumber. Spare no expense on your main posts, and be sure to check with local soil experts on the depth of hole you'll need for a solid placement. If you have brought in fill dirt, you may be required by local codes to dig down to undisturbed soil with each post, adding to your lumber expense.

Alternately, you may choose to surround each post with a "necklace" of concrete poured into the hole for security. Some builders feel a wide base is better than the necklace. Posts encased

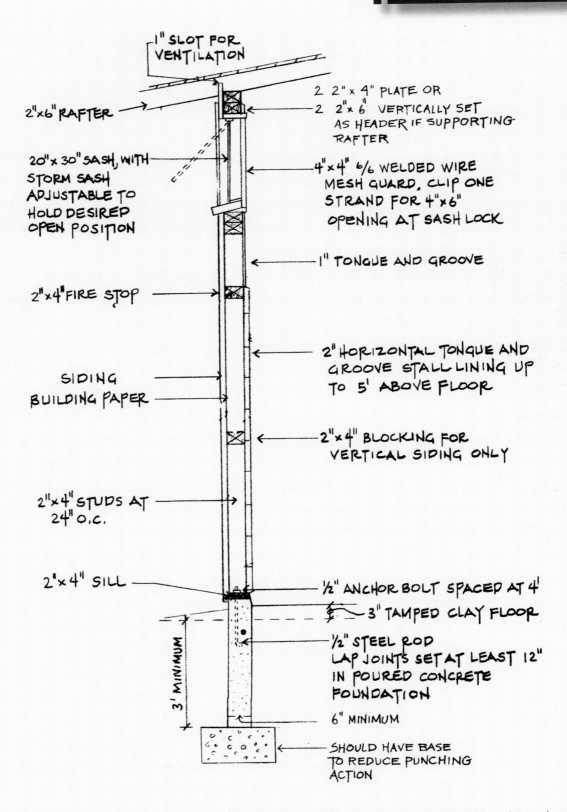

1" SLOT FOR VENTILATION

2"x6" RAFTER

20"x 30" SASH, WITH STORM SASH ADJUSTABLE TO HOLD DESIRED OPEN POSITION

2"x4" FIRE STOP

SIDING
BUILDING PAPER

2"x4" STUDS AT 24" O.C.

2"x4" SILL

3' MINIMUM

2 2"x4" PLATE OR
2 2"x6" VERTICALLY SET AS HEADER IF SUPPORTING RAFTER

4"x4" 6/6 WELDED WIRE MESH GUARD, CLIP ONE STRAND FOR 4"x6" OPENING AT SASH LOCK

1" TONGUE AND GROOVE

2" HORIZONTAL TONGUE AND GROOVE STALL LINING UP TO 5' ABOVE FLOOR

2"x4" BLOCKING FOR VERTICAL SIDING ONLY

½" ANCHOR BOLT SPACED AT 4'

3" TAMPED CLAY FLOOR

½" STEEL ROD LAP JOINTS SET AT LEAST 12" IN POURED CONCRETE FOUNDATION

6" MINIMUM

SHOULD HAVE BASE TO REDUCE PUNCHING ACTION

Here is a section of the wall in a barn built on a traditional foundation, that includes a window with a wire mesh cover.

Finishing the roof as soon as possible gives you a sheltered area to work under, which is invaluable in wet climates or if you build during the winter. (Joe Ann Scott, Fairfax Station, Virginia) Photo by N. W Ambrosiano.

in concrete for 15 to 20 years have a tendency to rot.

Before digging the post holes, mark the location for each along the string line. As the lines are marking the outer edge of the barn, you will need to locate the holes by measuring the width of the siding girt plus the center measure of your pole; allow 2" for the girt plus 3" for a 6" diameter pole. Accurate marking will save you from constructing walls that "wander" slightly and throw off your measurements for such standard items as plywood panels.

If you choose to build using 4" x 4"s, a handy spacing for them is every 4', allowing room for doorways and windows and plenty of support for the roof and loft. With posts 4' on center (o.c.), you always have a firm support for both hinges and latch on every door you install. Be sure, though, that if you make or order doors and plan to set them inside these posts, you take the 4" smaller inside dimension into account.

Using 6' and 8' pole increments is also effective if you use larger poles or

posts. These match standard lumber and siding units, which will save you milling costs.

Once the poles are in place, the splash board goes on—that is, the board that skirts the entire bottom edge of the building. Also known as the sill girt, it is the first of the girts that supports all of your siding, walls and rafters for the building. Use a 2" x 6" or 2" x 8" pressure-treated board here. If this first splash board is placed level, it represents a true horizontal for the rest of the building, and you can line up your siding and other work from this level board. You can pack stall flooring in any time after this treated board is in place.

With the pole and splash boards up, you can go directly to the roofing, if you'd like shelter as you complete the walls and interior. Or you can work from the ground up, framing in the stalls.

Exteriors

One of the most popular exterior sidings for modern barns is T1-11 plywood, an exterior grade of ply that has

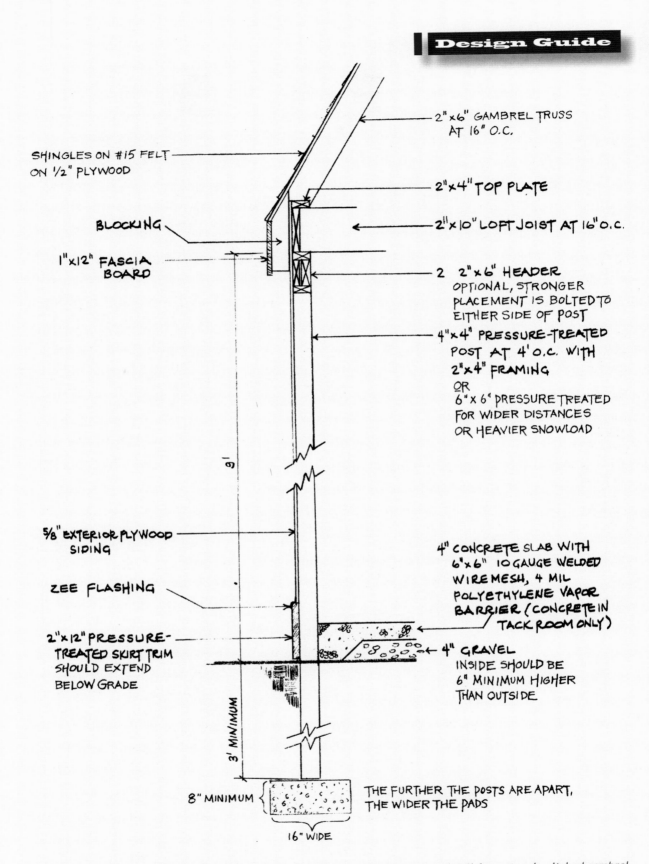

SHINGLES ON #15 FELT ON 1/2" PLYWOOD

BLOCKING

1"x12" FASCIA BOARD

2"x6" GAMBREL TRUSS AT 16" O.C.

2"x4" TOP PLATE

2"x10" LOFT JOIST AT 16" O.C.

2 2"x6" HEADER OPTIONAL, STRONGER PLACEMENT IS BOLTED TO EITHER SIDE OF POST

4"x4" PRESSURE-TREATED POST AT 4' O.C. WITH 2"x4" FRAMING OR 6"x6" PRESSURE TREATED FOR WIDER DISTANCES OR HEAVIER SNOWLOAD

9'

5/8" EXTERIOR PLYWOOD SIDING

ZEE FLASHING

2"x12" PRESSURE-TREATED SKIRT TRIM SHOULD EXTEND BELOW GRADE

4" CONCRETE SLAB WITH 6"x6" 10 GAUGE WELDED WIRE MESH, 4 MIL POLYETHYLENE VAPOR BARRIER (CONCRETE IN TACK ROOM ONLY)

4" GRAVEL INSIDE SHOULD BE 6" MINIMUM HIGHER THAN OUTSIDE

3' MINIMUM

8" MINIMUM

THE FURTHER THE POSTS ARE APART, THE WIDER THE PADS

16" WIDE

This section of a pole barn shows both the flooring for a tack room and the support detail for a steeply pitched gambrel roof.

a reverse board-and-batten appearance. It can be stained to match nearby buildings and comes in standard 4' x 8' dimensions, with options for grooves on 4", 8" or 12" centers. Install it vertically, and don't bother to cut out window holes before hanging it. It's easier to simply trim out the windows through the installed sheet with a hand-held power saw.

To join vertical sections of plywood, use special premade metal flashing if you want a weather-proof seal. Better yet, use a Plexiglas panel for the top 2' of a 10' wall, adding to your inside lighting.

A newer product is giving T1-11 some heavy competition in many areas. Called Smart Panel, it's expensive, but comes in a 9' length. It comes primed, ready to paint, and is water resistant. Its dimensions are not limited to plywood-type sheets, as it can be special-ordered in horizontal boards. (Smart Panel is a product of Louisiana Pacific. Contact information can be found in the Resources section.)

True board and batten makes a labor-intensive but handsome exterior. The boards are 1" x 6" or 1" x 8" lumber nailed vertically to the siding girts with narrow, 1- or 2", batten strips tacked over the cracks between them. As with any vertical siding, you must first apply horizontal nailers between the posts on which to tack the siding.

Tongue-in-groove, or shiplap, siding is another weather-tight siding that, if installed with the boards horizontal, lets you replace the ground level board as it weathers. Vertically placed sidings that rot at ground level must be replaced entirely, or patched along the bottom.

To avoid having to repaint every year or so, use wood stain on plywood or board siding. Oil stains last longer than paint, and, while they weather after a few years, they don't require painting or sanding before you apply the next coat.

Metal sidings are popular in some areas and can be extremely cheap and weather resistant. However, they do dent easily, so they must be securely lined against stray hooves and hips.

These metal sidings come in 8' lengths and 32" widths, but they can be special-ordered in any size you desire. Special fasteners—nails with washers built-in to prevent rattling and leakage—can be used as can screws. Check the type of nail or screw with the siding you buy, as you cannot use aluminum siding and galvanized nails. You can buy metal siding in a wide range of colors, which allows you to escape a large project—that of painting what you've built. If your structure is wood, add PVC or plastic washers between the metal siding and wood to prevent spot rusting.

Roofs

You have three main options in your roof construction: large timber beams with lumber decking, prefabricated wood trusses, or braced rafters. Local material costs and availability will have a strong bearing on the type you choose, but so will the amount of storage you wish overhead.

The most economical roofing system is often prefabricated trusses. These can be purchased from a fabricator, who will make them to fit your specifications. They preclude any overhead storage because of their many cross-webs and the fact that they are usually installed 24" center.

You or your builder can plan a stick-built system for your barn, which offers more storage capacity upstairs as it eliminates the diagonal webs needed in trusses.

No matter which roof system you choose, plan a 6- to 8" open space be-

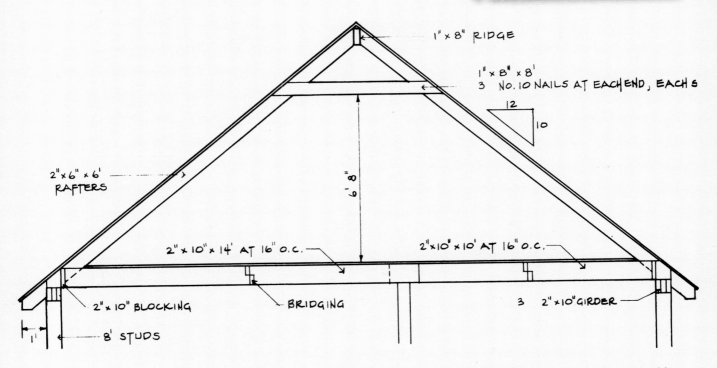

1" x 8" RIDGE

1" x 8" x 8'
3 NO. 10 NAILS AT EACH END, EACH S

2" x 6" x 6'
RAFTERS

6' 8"

2" x 10" x 14' AT 16" O.C.

2" x 10" x 10' AT 16" O.C.

2" x 10" BLOCKING

BRIDGING

3 2" x 10" GIRDER

8' STUDS

1'

A simple section of a gable roof includes a collar tie support on the rafters. Rafters are shown as 2" x 6"s but with wider barns or ones with large snow loads use 2" x 8"s.

neath the eaves to allow for good overhead ventilation. Should you wish to block this off occasionally, such as during a storm, you can install a hinged board at the top of the wall that can be swung up to close the gap.

Roofing Materials

You have a wide range of roofing material options, depending on the appearance you desire. In the Southwest, handsome clay tiles make a cool, waterproof roof, blending well with the local construction style. In Eastern suburban neighborhoods, fiberglass shingles help the barn fit in, and provide a good, durable surface. For the cheapest roof, metal, whether corrugated steel or aluminum, is your answer. It can be noisy in rainstorms, but it is extremely easy to install and requires no maintenance. If you live on a coastline, where salt breezes wreak

havoc on regular steel roofs, aluminum is your best choice. Remember to use only aluminum nails with this material, or the roof will react chemically and become damaged.

If a top-quality metal roof fits your design ideas, several companies produce roofs of every shape, color and level of strength. Known primarily for their industrial work, such as on warehouses, office buildings and restaurants, they can cover your barn with roof material that will hold up long after you and your horses have moved on to greener pastures. These top-quality metal coverings are made in a rainbow of colors, often with custom shades available, and they can be lined to reduce noise and condensation. Metal roof manufacturers also provide metal sandwich-panel roofing, with built-in rigid insulation, cutting down on noise and heat to the

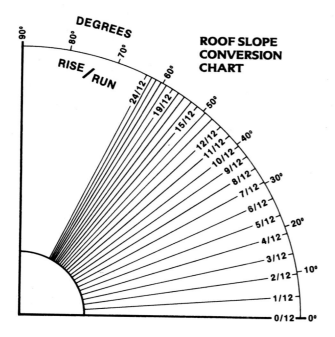

DEGREES

RISE / RUN

ROOF SLOPE CONVERSION CHART

90°
80°
70°
60°
50°
40°
30°
20°
10°
0°

24/12
19/12
15/12
12/12
11/12
10/12
9/12
8/12
7/12
6/12
5/12
4/12
3/12
2/12
1/12
0/12

outside. Or, shed insulation can be put in place under the metal roofing to cut down heat radiation and afford some sound insulation.

A metal roof is installed over a simple grid of nailers between the rafters, giving you the fastest shelter with the fewest materials. It requires no heavy decking or underlayment, and has a long lifespan.

Roof Loads and Support

If you are unsure just how many rafters, trusses or support poles you need to both anchor and support your roof, it's best to check with a local building supply company, your state university's engineering department, or private engineers. Ask about the loads, and abilities your barn will require to stand up to your local weather conditions. Besides just holding up the roof and shingles, your roof must be able to withstand wind and snow common to your area. All of the elements in a barn need to work together to resist all of the loads applied on it, and due to the variability of the loads, this book cannot convey all of the structured information needed to build a safe barn.

The national Building Officials & Code Administration is another source for recommended load-bearing characteristics your roof should have. Check your local library, the Internet or with your local planning department, for help on this. Don't forget to check with building professionals in your area, even if you're not using them for the whole project.

You also need to understand the load-bearing characteristics of the lumber you plan to use. The size and stress rating of the wood are key factors in its ability to both support your load and withstand the weather. If you want to deviate from your plan on the recommended type and size of lumber, be sure to ask your supplier or builder if the substitution is in the best interests of your building. Unlike house construction, where some walls are bearing—they support the weight of the structures above—most of the barns in this book are pole barns, where the structure basically hangs from the poles you set at the corners and along the walls. That gives you a great deal of flexibility in your wall materials throughout. Your roof, however, must be strong enough to withstand upward blasts of big storm winds—even hurricanes, in some cases—and at the same time stay in place with a possible load of snow on top.

Most roof load capacities are calculated in pounds per square foot. The Midwest Plan Service (see Resources list) has a chart detailing the loads and recommended supports for them. You can then figure the pounds your roof can handle, based on your lumber's F-number, for the pounds of pressure it can withstand.

Lumber can be machine tested to determine its stress capacity, giving the purchaser a standard by which to judge the strength of the building's materials

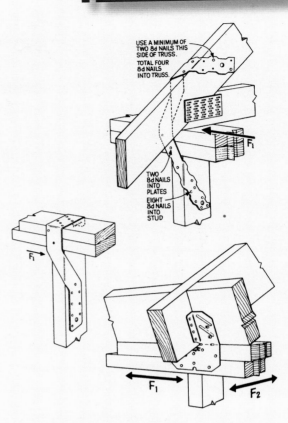

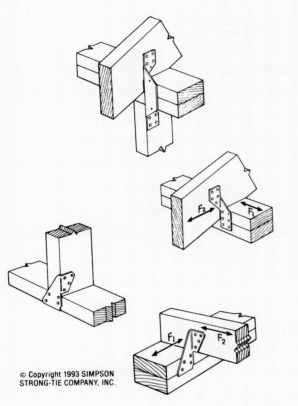

© Copyright 1993 SIMPSON STRONG-TIE COMPANY, INC.

A selection of carefully engineered ties such as these will go a long way toward securing your barn in high winds or earthquakes. Chosen and placed according to the type of application needed (to tie rafters against uplifting winds, or to secure framing members to bottom plates or sill girts, etc.), they help your barn hold together under stress, as the connections between wooden members are what will come apart first. The group shown here, from Simpson Strong Tie Company, Inc. (contact information located in the Resources section) is designed specifically for hurricane and seismic load, and many parts come sized for either finished or rough-sized lumber.

(provided, of course, the hardware is up to the same challenge as the lumber). Most lumber used for barns is "eye graded" rather than machine graded. This is adequate for most users for barn construction. If you have an area that needs particular strength requirements, such as a section where a home or apartment will be built overhead, machine graded material, though more expensive, might suit your project better.

Durable shingles will last up to 25 years and can be ordered in colors that complement nearby buildings. However, they do require considerable preparation: You must lay a ½" plywood deck over the rafters, a layer of roofing felt, and finally the shingles themselves. Instructions for installation vary with the brand of shingle, and are usually included on the package.

Regardless of the roofing material you choose, plan to install gutters on your barn. Tacked on to the fascia boards on the ends of the rafters, gutters will keep your barn from disappearing behind a sheet of water every time it rains. Vinyl or aluminum gutters are inexpensive, simple to install, and adequate in all areas but those of high snowfall. If you have a steeply pitched roof and severe snow, ask local building specialists for their recommendations. You may need to resort to reinforced guttering,

With a manufactured barn, such as the Barnmaster, Inc., facility shown in these photos, you can look forward to a swift completion of your project once the truck rolls into the yard. (Photo courtesy of Barnmaster, Inc.)

With a previously prepared site, such as this one, all sanded, packed and set for walls, a sharp crew can erect the pre-made panels and connect them in a matter of hours. Expansions and changes are simple as well, since the components are consistent throughout the project. (Photo courtesy of Barnmaster, Inc.)

brackets to hold back snow, or other options. You must keep the gutters clear to prevent roof damage.

If you choose to go without gutters, you can run a grade strip of gravel away from the barn that will catch the falling water. First, dig a trench along the drip line, leading to a drain field, and then fill it with gravel and edge it with treated scrap lumber. The fall should be at least $\frac{1}{8}$" per foot.

WORKING WITH MANUFACTURED-BARN COMPANIES

Not all barns have to be built on your property, piece by piece and board by board. An option is to order a package from one of several manufactured-barn companies and have the sections assembled on a prepared slab.

This is certainly one of the fastest ways to establish a facility, and there are fewer nagging design questions to haunt you. Those details are sorted out by the corporate design-and-construction folks,

leaving you to choose the package you like. A modular barn is not right for everyone, though, because of price, style, climate variations and availability. If you choose to go with a manufactured barn, it pays to closely examine as many as you can find in your area, four seasons of the year. Be sure the company you are considering can recommend a nearby example of their own work.

Choosing Your Company

Manufacturers of modular barns are located around the country, giving you a variety of choices in style, components, and shipping costs. Horse owners in the Western states have an advantage when it comes to shipping costs as metal-frame modular barn manufacturers are extremely common there. These companies offer everything from the simplest two-stall configuration to gorgeous showcase facilities with scores of stalls, attached indoor arenas and living facilities to boot.

The popularity of manufactured barns can be attributed to the benefits of fire safety and reduced lumber costs. The metal framing is often combined with metal-covered wood walls,

which are all but fireproof, according to manufacturers. While strong and indeed fire-resistant, some metal-covered walls have been known to "sweat" a bit in some weather, and they can become less attractive (although will hold up) in the vicinity of a kicking horse.

Alternate materials are available from manufacturers, and whatever you choose will depend on your funding and preferences. Explore the possibilities of interiors that vary from those shown in the brochures or in your local area. Whether your preference is tongue-in-groove wood, foam-insulated or galvanized steel walls, you can make arrangements for any of these.

No matter which company you choose, talk to the owners of that company's barn projects and look over their materials closely, as you would with any other construction project. Your horses' energy level, size and manners in the barn may call for moderate to extremely strong materials and connectors, items that will vary from one company to another. And an item as seemingly insubstantial as a weld or a hinge may make all the difference between a safe horse and an escapee.

Planning Your Barn's Delivery

In most cases, manufactured-barn companies have local representatives who can work with you in person or over the phone to establish your needs. These representatives will ask you about your general ideas and needs, then send sales materials about their current products so you can see their designs in full color, complete with floor and roof plans.

Follow-up calls and faxes will provide sketches and cost estimates, allowing you to develop a solid idea of what will surround your horses should you go ahead with the project. Have a few things in mind as you prepare to make book:

■ Know your budget and basic needs, but don't be rigid; these folks are professionals, with many barns in their backgrounds. They might just have an answer you haven't thought of. At the same time, they can't read your mind, so be clear and detailed in your requests.

■ Meet with the builder's representative and see what suggestions he or she has for your needs, then

The final product (right), one of Barnmaster's two-stall, shedrow models, represents a straightforward, workmanlike setup in the western U.S. (Photo courtesy of Barnmaster, Inc.)

Built to house breeding stallions, this structure at Gainesway Farm has four stalls, a center aisle and a work/storage area. The arched, vaulted ceilings anchored into the ground and thick concrete walls were built to give as much protection as possible from tornados and related high winds. (Gainesway Farm, Lexington, Kentucky) Photo by Stephanie Church.

expand your wish list to include available extras, such as hay doors, roof fans, and stall accessories, as your budget allows.

■ Confirm that planned-for items, structural detail and even colors conform to local zoning, building and neighborhood codes. Your company's representative may know requirements for your county, but don't bet on it.

■ Finalize plans, colors, placement, etc., as well as your expected delivery/completion dates.

■ Remember that in most cases, you must prepare your own site. You'll need to arrange for concrete, gravel, water, electricity and tele- phones. Be sure to consider these costs in comparing your modular project to a site-built facility so your comparisons are valid.

Handling Weather Extremes

Let's be realistic: No matter what you build, it's not going to withstand the worst of what nature can throw at it. But you can do a reasonable amount of planning to minimize the trauma to your facility and your horses in the case of tornados, hurricanes, grass fires or earthquakes.

A direct hit from a tornado is going to cause damage to your place, but you can escape serious harm from the near-misses by strapping down your roofs, and bolting or firmly sinking your support posts. Use some of the new attachment hardware available for securing posts, joists and rafters so that they are linked firmly through each other to the ground. Don't depend on gravity to hold your place together at any point. Use roofing materials such as fiberglass shingles that don't catch the wind as much as other materials, and that won't be lethal if they become airborne.

Gainesway Farm in Lexington, Kentucky, is a fine example of a farm that has constructed buildings for their stallions designed to minimize the chances of major losses given a direct hit from a tornado. This high-profile stallion station stands some of the most valuable horses in the world, so a major investment was called for.

First, the owners of Gainesway limited their barns to four-stall units, so the whole group of stalls can't be wiped out at once. Their arched roofs run diagonally from one corner to another and are anchored deep into the ground with concrete. The walls are l' thick poured concrete, and their windows are narrow rectangles placed high under the eaves to protect against flying debris. These are highly specialized barns.

Hurricanes are much like tornados in terms of high winds that tear at your roof and peel off anything that can catch a breeze. News coverage of hurricane damage in Florida has shown horrific injuries to horses who'd been stabled under or near corrugated metal roofs. (The metal flew off almost immediately as the winds grew, and it cut nearby horses like knives as it blew around the stalls, aisles and pastures.)

Fortunately, you usually have some warning of a hurricane's approach, and you should be quick to evacuate your animals to safer areas. In making this choice, look over not only your facility's design, but that of nearby properties as well; your neighbor's roofing material may become your problem.

In an earthquake, your barn is at the mercy of the ground it's built on, but it's likely to fare fairly well, given the inherent flexibility of a pole barn. Your soil type, as well as the distance from the epicenter of the quake, will define the level of shaking you receive. Bedrock and native, layered soils are the most solid, seldom amplifying the quake's effects. Manmade fill, on the other hand, shakes like jelly when the ground moves, and it can become almost liquid in its motion, depending on the type and duration of the shock waves.

In general, if you are in an area with any known quake zones, use the same precautions that you would in your home: Secure tall cabinets to the walls, keep them latched closed, place nothing above shoulder level that would hurt you (or your horse) if it fell on you, and have plenty of extra water on hand in case the quake ruptures your water lines or contaminates your water supply.

A note of experience from the 1989 Loma Prieta quake in Northern California: Don't rush to turn your horses out, as they'll hurt themselves by bolting through your fence lines. Unless your roof really is coming down, stay safely under a door frame or ridge pole and avoid possible chaos outside. If you have trees overhanging your barn, check them for dead branches and accumulated debris each year, as your roof may be damaged and horses frightened by the falling tree materials.

READING BLUEPRINTS

It takes architects and draftsmen several years to become proficient at designing buildings and producing the drawings they are built from. Reproduced drawings are known as blueprints. While you can't become an expert in reading blueprints from just a few paragraphs on the subject, you can begin to understand what they can tell you and what kinds of questions you may need to ask your builder.

Scale

Most plans are drawn to scale. A legend or key in a bottom corner should tell you the scale of the drawing. This scale will aid you in determining the relationship of the buildings on the land for a site plan and the interior layout of walls, and electric and water lines on the floor plan. Any other details pertaining to the construction of your facility can also be included in the blueprints you've had prepared or the basic drawings you did yourself.

Site Plans

A site plan shows the land, its topography and the arrangement on the land of the many buildings, fencing, driveways, riding rings, etc., you've designated for your facility. This plan should be one of the first decisions you make after purchasing your property and deciding what you need or want to build. This will be a work in progress as you walk the property, planning the flow of your operation and the day to day work involved in running the facility. This plan many change as you analyze information on soil conditions, weather patterns, drainage and the buildings you will make part of your facility. Once you've decided on where everything will go, you can concentrate on the drawings for your buildings.

Floor Plans or Working Plans

The floor plans, also called working plans, are the agreement between you, the architect (if you used one), and the builder as to what will be built, the size of the materials, and the overall look of the finished product. Since the working plans constitute part of the contractual agreement between all of the involved parties, it's important you examine them carefully, understand the rudiments of their meaning and be certain the plan fits your needs. Understand that changes to the plans may be necessary during the construction phase. Last minute changes mean the builder and workers must rearrange their plans and scheduled work to accommodate your changes. While you should never hesitate to make changes in your original plans if the need arises, a good working relationship with your builder will help to smooth the process should any changes become necessary. Find out what the builder's schedule will be for your project, and understand the builder doesn't want a crew standing around while concrete dries before they can continue building. You don't want to pay for that down time, either. A blueprint can aid in all this by helping you understand the sequence of construction. After all, the foundation must be built before the walls can be put in place.

Your blueprint may contain several pages of drawings for each building. In addition to the general floor plan layout, there may be structural drawings of the roof, and mechanical, electrical and detail drawings of any special support members like headers for large barn doors on the ends. Cut away drawings, known as sections, show details that will be hidden behind walls, floors or ceilings.

Elevation Drawings

Elevation drawings show how high the buildings will be and the shape of the roofline when viewed from the side. Elevation drawings can also show items such as chimneys, porches, windows, doors or any other exterior material that will become a permanent part of the building.

Sectional Drawings

Usually larger in scale than the other parts of the drawings, sectional drawings will show a cut-away schematic of constructional components that pay special attention to construction details.

Special Plans

Details for electrical, plumbing, heating, air conditioning or any other specific layouts of internal components can either be included in the main floor plan or be drawn as an auxiliary plan. The more complicated your construction, the more plans necessary so that the builder(s) can interpret them and construct your building correctly. You may want to use several contractors. If you are going to be the general contractor and subcontract the jobs out to a variety of contractors, you may save a lot of money. Being the general contractor means you are not paying someone else to take full responsibility for the entire job. It also means that you can choose the best contractor for any unique or specific jobs. The builder you use may appreciate this. While your builder may

do a great job in many areas, installing something as specific as a garage door can be done quicker, easier and cheaper by the specialty company that installs these doors all the time.

Revisions

Should you wish to make changes to your original plans, the best time to do it is before construction begins, when revisions can be incorporated into the original drawings. Be sure that any revisions are duly noted and only the most current changes are used during construction so that the revisions are incorporated as building continues, to avoid mistakes.

If any changes need to be made during construction, be certain that all parties completely understand what the changes will be and what the associated cost will be as a result of the changes.

Final Landscaping Plans

After all construction is completed you will want to landscape certain parts of the property. This is different from the pasture and horse-related grading and seeding that you will have done as part of your overall farm plan. Landscaping construction will grade soil away from buildings to improve drainage and add beauty to the overall facility design. A landscape architect may be needed for this part of the project.

In conclusion, architects, draftsmen and builders use a blueprint to guide and direct their design and construction of buildings. The structure must support the weight of all construction materials as well as withstand weather normal to the area and movement associated with the day to day operation of doors and windows. Getting the blueprint right at the onset of construction will save you time and money as the facility is built.

Barn Plans

½" PLYWOO
WITH 6 THR

2" x 6" BRAC
HORIZONTAL
IN SNOWLOAD

5'

3' 10'

2"x10"x10' AT 16" O.C.

3   2"x10" GIRDER

BRIDGING

he barns shown in this chapter solve a particular problem or horsekeeping challenge for each of their owners. They should thus be viewed in the context of their locations and their owners' needs. The dimensions and lumber types provided for each are offered for conceptual purposes only, as actual sizing of elements and layout for your own facility may need determination by a designer or engineer fluent in your local conditions and codes.

This variation on the run-in shed allows for three sides to be closed for more wind-rain protection, and yet the high roof and wide open area on top of the walls allows for heat to rise and winds to dispel it. Here the hay rack is located on the side wall, giving easier access to those refilling it and providing a safe location in which horses can eat. Placed to the rear of a small shed, a hay rack can inspire one horse to keep others at bay with his hindquarters as he eats. If the rack is on the side, as shown here, he is less likely to dominate the entire shed. (Hassell Arabian Stud, Reddick, Florida) Photo by M. F. Harcourt.

FLORIDA SUN SHED

Description

A run-in shed is the ideal arrangement for horses in all but the most severe climates. The animals are allowed shelter, but not enclosed where they can develop leg and lung problems. In Florida, as in many parts of the country, a shed need only offer shade from the sun; wind and snow are not an issue. Thus, this shed at Hassell Arabian Stud in Reddick, Florida, has only a slight pitch to its roof, serving as little more than a permanent parasol and hay feeding area.

MATERIALS

Posts:
   6" x 6" pressure-treated or creosoted

Rafters:
   2" x 8" nailed with 16d nails

Framing:
   Add braces for wind up-lift 2" x 8" joists

Roof:
   Sheet metal with nails or same metal

A close-up shot shows a hay rack with a trough below it that catches fallen leaves from legume hays and allows for the feeding of salt, minerals, supplements or grain. Photo by M. F. Harcourt.

Paul Hassell has found a way to provide some protection for his horses from the Florida sun by erecting these simple open run-in sheds. Since the climate is rarely severe, they function as year-round shelter for both his horses and his cattle. Photo by M. F. Harcourt.

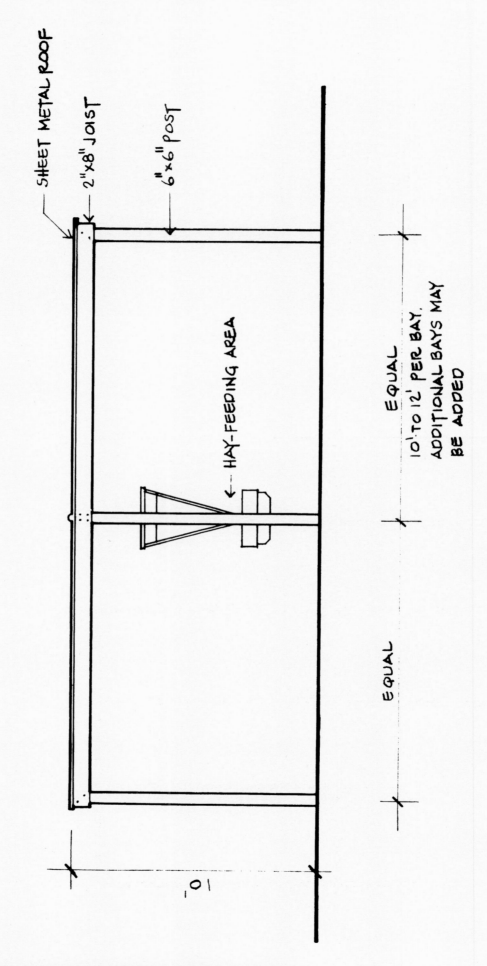

SHEET METAL ROOF

2"x8" JOIST

6"x6" POST

HAY-FEEDING AREA

EQUAL

10' TO 12' PER BAY.
ADDITIONAL BAYS MAY
BE ADDED

EQUAL

10'

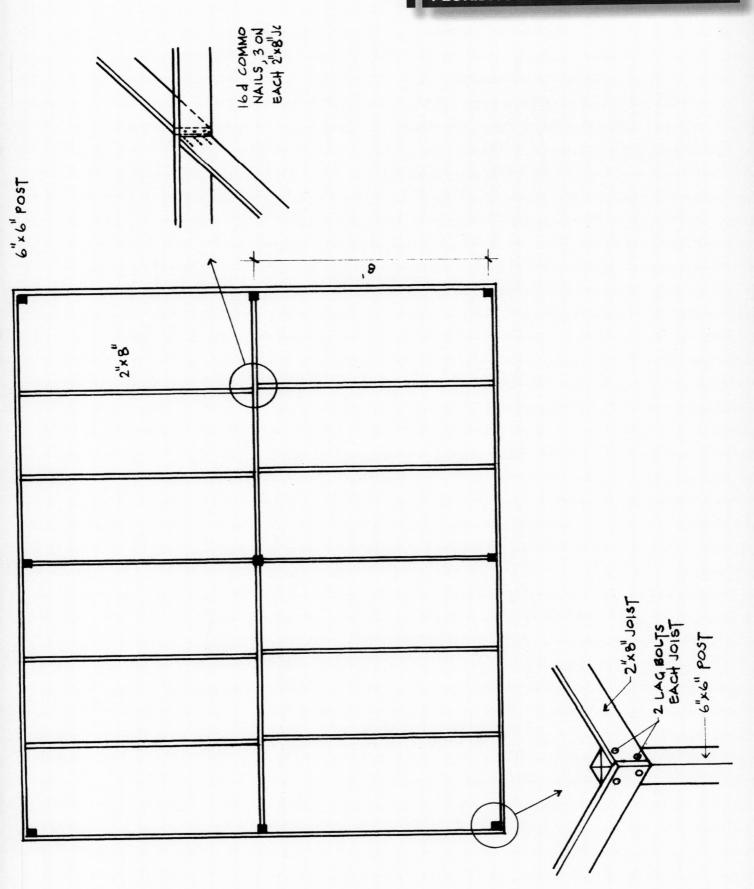

16d COMMO
NAILS, 3 ON
EACH 2"x8" Jc

6"x6" POST

2"x8"

8'

2"x8" JOIST

2 LAG BOLTS
EACH JOIST

6"x6" POST

Each 28' roof in these shed combinations shelters a 12' x 12' area for each horse, plus a 4' x 12' storage area between them. A 4" x 4" beam along the sill of the shelter keeps bedding from being trailed out into the paddock. The support posts are 6" x 6" timbers, while the roof is of principal rafter and purlin construction. Fiberglass reinforced plastic or corrugated metal may be used for roofing. Photo by N. W. Ambrosiano

CALIFORNIA RUN-IN SHED

Description

This shed arrangement in Nicasio, California, at Carmen Johnson's Windfield Station, is ideal in an area with paddocks for individual horses, or it can be set over the junction of two full pastures. There is horseproof storage in the center aisle, and the sides can be enclosed as snugly as your climate requires.

To keep a horse in overnight, such as before a show or hunt, one could use a 2" x 8" board that drops into place across the main opening. To secure it, make a small channel of 2" x 2" at your horse's chest height that each end of the board will drop into. Or, nail a horseshoe to each corner support across the front, with the shoe extending halfway out from the post. The board will slip into the hook provided.

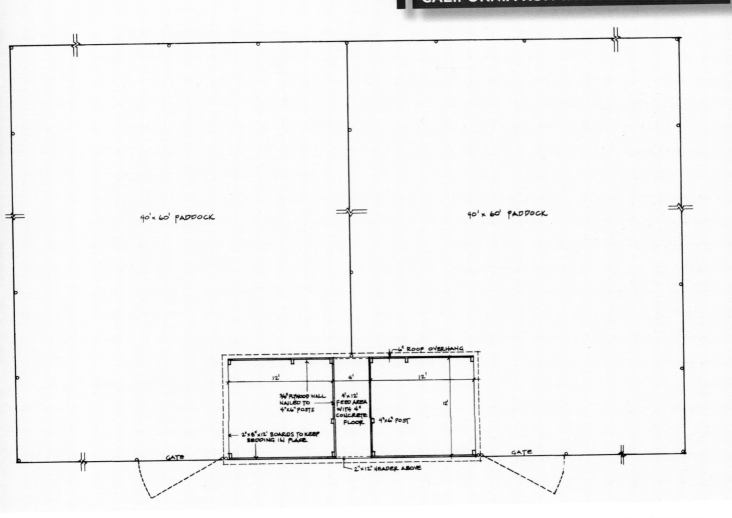

40'x 60' PADDOCK

40'x 60' PADDOCK

6" ROOF OVERHANG

12'

4'

12'

3/4" PLYWOOD WALL NAILED TO 4"x6" POSTS

4'x12' FEED AREA WITH 4" CONCRETE FLOOR

12'

2"x8"x12' BOARDS TO KEEP BEDDING IN PLACE

4"x6" POST

GATE

GATE

2"x12" HEADER ABOVE

In this California boarding farm, a row of paddocks and run-in sheds allows easy access for each owner and enough shelter for horses in the mild climate. Each shed is centered over the fence between two paddocks, allowing one roof to serve two sheds, with a small storage area between them. Photo by N. W. Ambrosiano

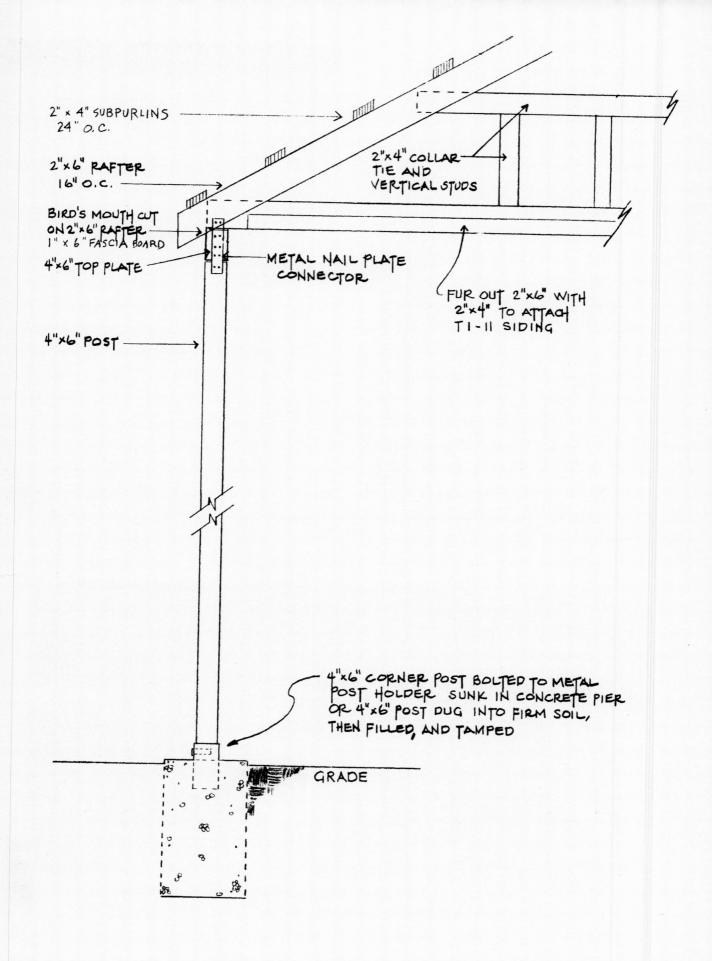

2" x 4" SUBPURLINS
24" O.C.

2"x 6" RAFTER
16" O.C.

BIRD'S MOUTH CUT
ON 2"x 6" RAFTER
1" x 6" FASCIA BOARD

4"x 6" TOP PLATE

4"x 6" POST

2"x 4" COLLAR
TIE AND
VERTICAL STUDS

METAL NAIL PLATE
CONNECTOR

FUR OUT 2"x 6" WITH
2"x 4" TO ATTACH
T1-11 SIDING

4"x 6" CORNER POST BOLTED TO METAL
POST HOLDER SUNK IN CONCRETE PIER
OR 4"x 6" POST DUG INTO FIRM SOIL,
THEN FILLED, AND TAMPED

GRADE

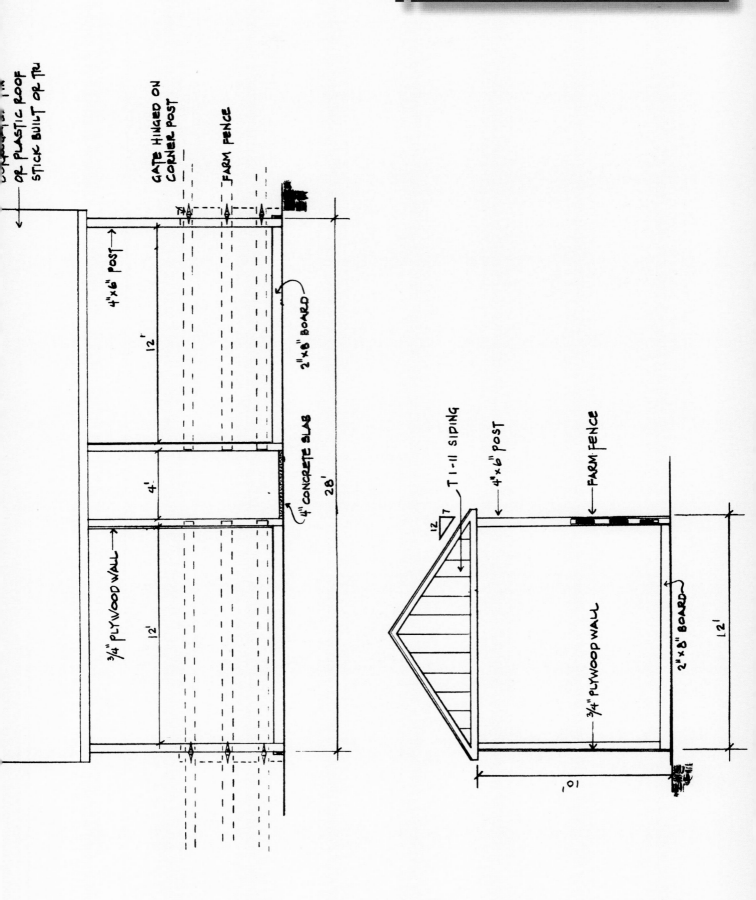

CORRUGATED TIN OR PLASTIC ROOF STICK BUILT OR TR

GATE HINGED ON CORNER POST

FARM FENCE

4"×6" POST

12'

2"×6" BOARD

4' CONCRETE SLAB

4"

28'

3/4" PLYWOOD WALL

12'

12'

T 1-11 SIDING

12 7

4"×6" POST

FARM FENCE

3/4" PLYWOOD WALL

2"×8" BOARD

12'

10'

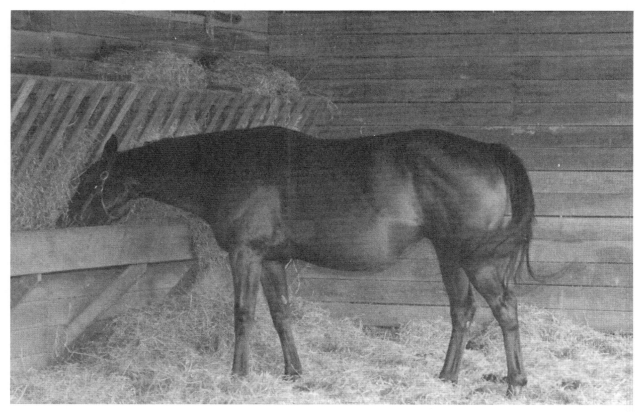

Note the solid double-wall construction here that may not be necessary in warmer climates or conditions. The double walls with studs add to the cost but keep horses from kicking loose and injuring themselves on boards that are nailed only from the outside. The minimum shed depth of 20' allows the horses to move deep enough in the interior to get away from the flies in summer and the biting winds of winter. The length of the facility can be in any 12' increment that will satisfy your needs. Some sheds are constructed with horse-proof space behind the manger to store hay or other feeds as needed for ease of feeding in any weather. (Bittersweet Farms, Middleburg, Virginia) Photo by M. F. Harcourt.

VIRGINIA RUN-IN SHED

Description

In Virginia, run-in sheds must not only provide protection from the summer heat but also be a buffer from frigid winter winds. Some farms have used run-ins for broodmares, mares and foals, and yearlings. They have found fewer upper respiratory and injury problems with stock, and labor requirements are reduced.

MATERIALS

Poles:
  6" x 6" pressure-treated

Framing:
  Sill girts 3 courses 2" x 6" tongue-in-groove (t&g) pressure-treated lumber

  Siding girts 2" x 4"
  Rafter girts 2" x 6"

Rafters 2" x 6" 2' o.c.

Exterior:
  T1-11 siding

Interior:
  1" x 6" rough-cut lumber

Roofing:
  ½" CDX plywood
  15 lb. felt
  215 lb. asphalt shingles

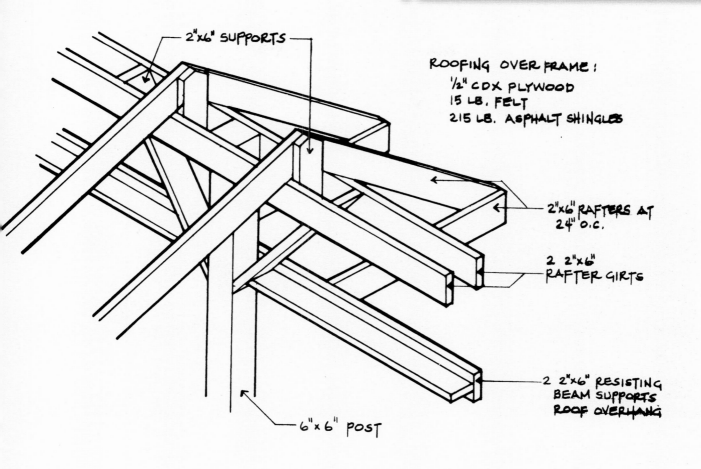

2"x6" SUPPORTS

ROOFING OVER FRAME:
½" CDX PLYWOOD
15 LB. FELT
215 LB. ASPHALT SHINGLES

2"x6" RAFTERS AT 24" O.C.

2 2"x6" RAFTER GIRTS

2 2"x6" RESISTING BEAM SUPPORTS ROOF OVERHANG

6"x6" POST

This view of the overhang in the run-in shed shows how it ties in to the roofline at the doubled ridge girt and then is supported by horizontal braces nailed to a second girt farther down the posts. A second board, nailed perpendicular to the lower girt, strengthens the girt against the lateral thrust of the overhang supports. (Bittersweet Farm, Middleburg, Virginia) Photo by M. F. Harcourt.

These full views show how the run-in shed can be used as a primary source of shelter. (Bittersweet Farm, Middleburg, Virginia) Photo by M.F Harcourt.

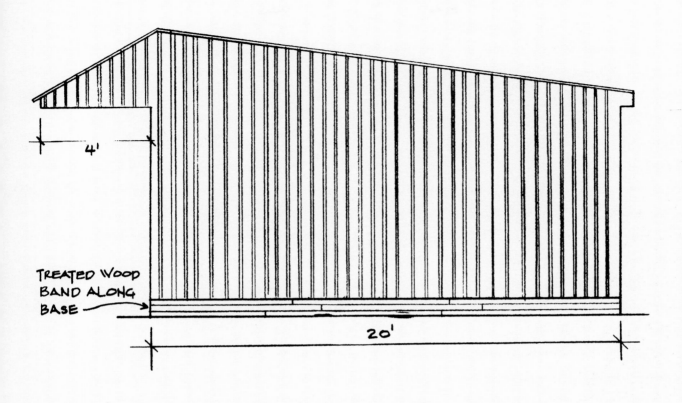

4'

TREATED WOOD
BAND ALONG
BASE

20'

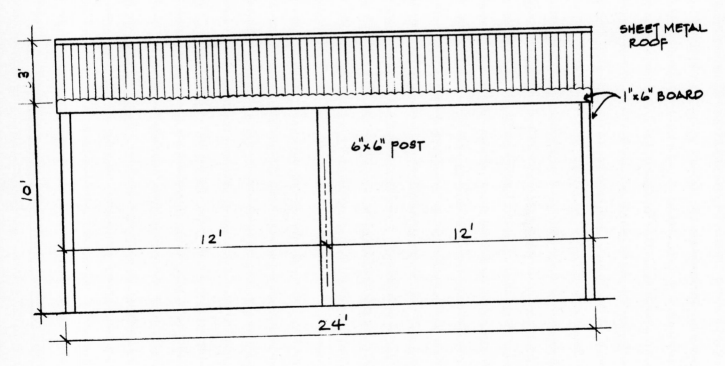

SHEET METAL
ROOF

3'

1"x6" BOARD

6"x6" POST

10'

12'

12'

24'

ONE- OR TWO-HORSE BARN

Description

The barn-building company of P. J. Williams, Inc., of Somerset, Virginia, designed this little barn, which is an ideal arrangement for those with a horse or two and minimal equipment. This can be built with several options, such as overhead storage, an open roof area, or a gambrel roof, as opposed to a simple gable roof. With the poles spaced as they are, one of the stall doors could be widened to allow more storage space if the stall wasn't needed for a horse. In addition, pull-down steps to a loft could be installed in the tack and feed area.

For warmer climates that call for more ventilation than the combined door and window area provides, an eave vent could be installed, as well as additional side windows.

MATERIALS

Dimensions:
  17' 9" x 31' 9"

Poles :
  4" x 6" pressure-treated at the base

Framing:
  Rafter plates 2" x 10"

  Rafters 2" x 6" 2' o.c. with ½" ply gussets on gambrel model

  Optional loft joists 2" x 10" 16" o.c.

  Sill girt/skirt board 2" x 6" tongue-in-groove pressure-treated lumber

  Siding girts 2" x 4"

Roofing:
  ½" CDX plywood

  15 lb. felt

  215 lb. shingles

Siding:
  T1-11 textured plywood

Doors :
  Two 4' x 8' Dutch door sets

  Two standard exterior-grade house doors with diagonal trim as desired

Windows :
  Two 3' x 4' single-hung aluminum frame or a screen of vertical bars installed 3" o.c.

NOTE:
IN AREAS WITH SNOWLOAD,
CHECK ALL ROOF DESIGNS
W/ENGINEERS FOR ADEQUATE
WEIGHT-BEARING SUPPORT

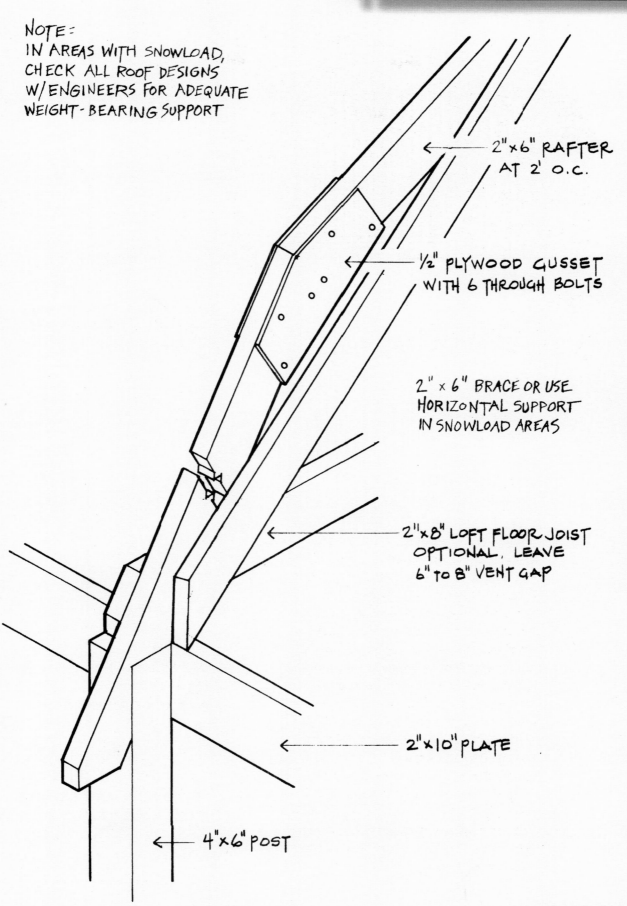

2"x6" RAFTER
AT 2' O.C.

½" PLYWOOD GUSSET
WITH 6 THROUGH BOLTS

2" x 6" BRACE OR USE
HORIZONTAL SUPPORT
IN SNOWLOAD AREAS

2"x8" LOFT FLOOR JOIST
OPTIONAL, LEAVE
6" TO 8" VENT GAP

2"x10" PLATE

4"x6" POST

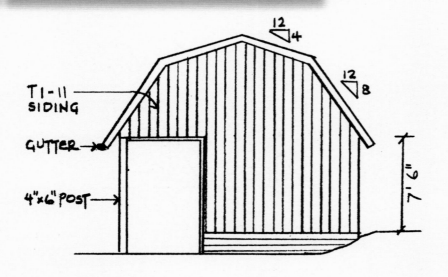

T1-11
SIDING

GUTTER

4"x6" POST

12
4

12
8

7' 6"

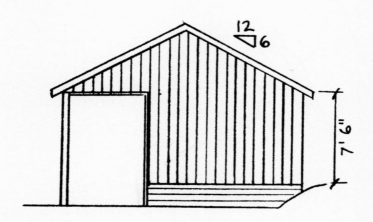

12
6

7' 6"

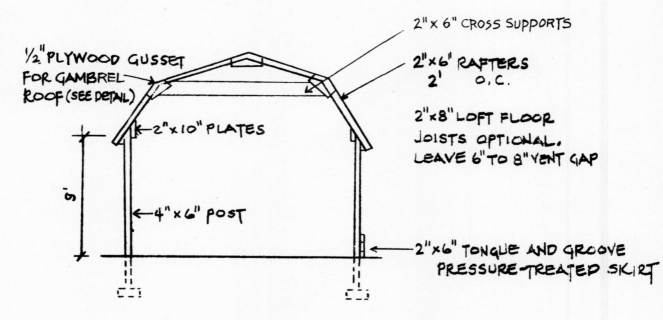

½" PLYWOOD GUSSET
FOR GAMBREL
ROOF (SEE DETAIL)

2"x10" PLATES

9'

4"x6" POST

2"x 6" CROSS SUPPORTS

2"x6" RAFTERS
2'   O.C.

2"x8" LOFT FLOOR
JOISTS OPTIONAL.
LEAVE 6" TO 8" VENT GAP

2"x6" TONGUE AND GROOVE
PRESSURE-TREATED SKIRT

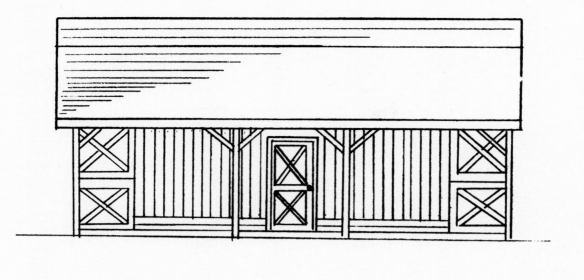

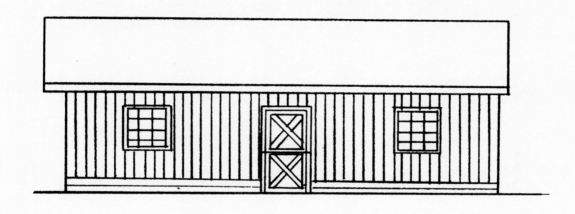

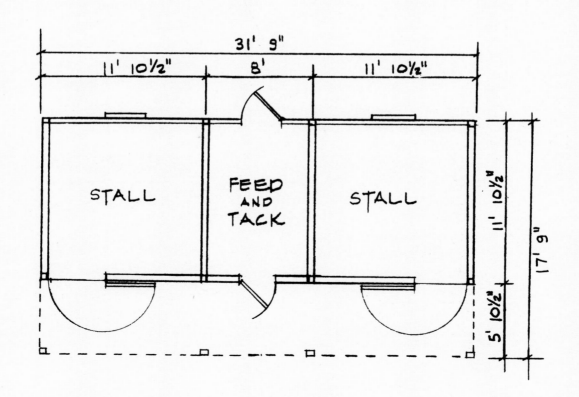

The simplest of arrangements have been made here, such as a central water supply with a hose to reach into the stalls.

THREE OR FOUR-HORSE SIMPLE BARN

Description

This simple, workmanlike barn is one that can be built by an inexperienced person with a minimum of trouble. With no overhead hay storage, it provides plenty of airflow through the rafters, its simple pole construction allows for varied ground slope around it, and it can be added to with a minimum of trouble. In its simplest form, it is just stall spaces with an overhang. A 6' x 8' storage area can easily be added at the left end of the aisle where it won't interfere with access to any stalls. While this version, built by Helen Makarov of Middleburg, Virginia, is shown with board and batten siding, any other can be substituted with only minor modification of the horizontal nailers between the girts.

This simple 20' x 40' barn can be used anywhere you need the basics of shelter, even if your site is uneven and you choose not to fill extensively. This barn, built by Helen Makarov of Middleburg, Virginia, has been built into the sloping site. While the roofline is level, the stalls are stepped down the hill. As a result, the left and right outside walls are not the same height, resulting in varied material requirements. Note the two 2" x 10" ridge support beams are well supported, precluding the need for thrust ties. Photo by M. F. Harcourt.

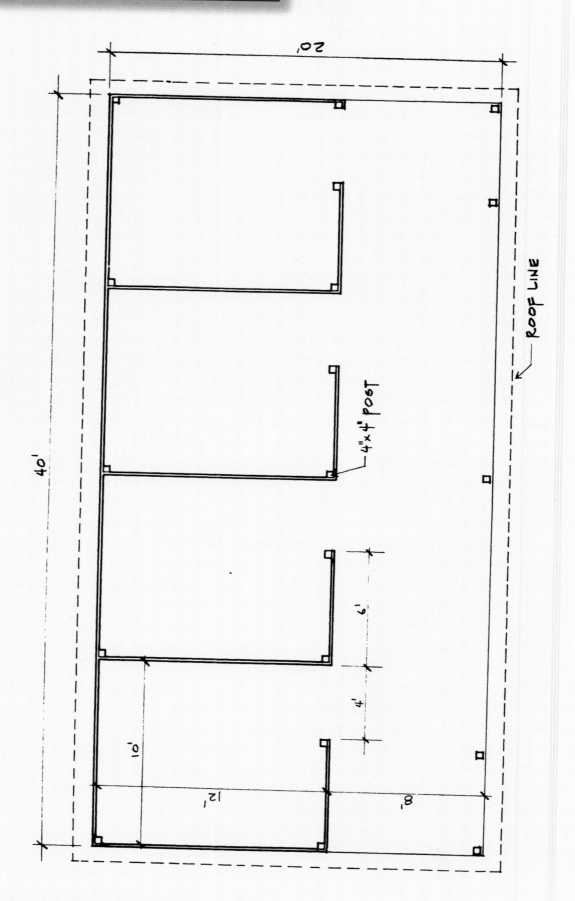

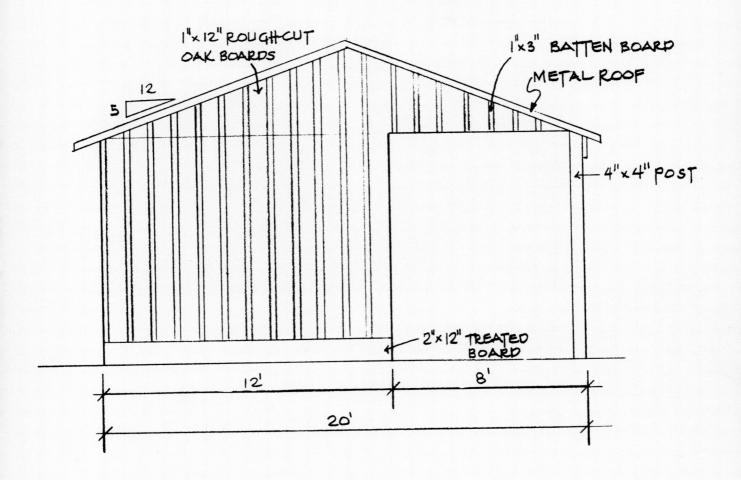

1"x 12" ROUGH-CUT OAK BOARDS

1"x 3" BATTEN BOARD

METAL ROOF

5 | 12

4"x 4" POST

2"x 12" TREATED BOARD

12'

8'

20'

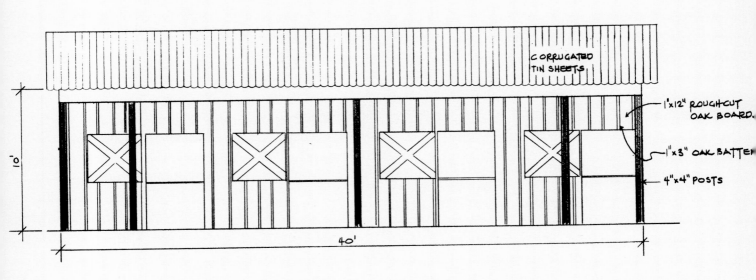

CORRUGATED TIN SHEETS

1"x 12" ROUGHCUT OAK BOARD.

1"x 3" OAK BATTEN

4"x 4" POSTS

10'

40'

Utilizing land contour to their advantage, the owners of this barn built into the side of a slope in Colorado have taken advantage of terrain to offer both weather protection and loft access. The key is the concrete retaining wall buttressed up against the slope. Planning for water run-off and drainage around this barn is imperative in an area with anything but an arid climate.

COLORADO BANK BARN

Description

This simple, extremely serviceable barn is built into the side of a hill in a design strongly reminiscent of early New England and Pennsylvania barns, although its finish style reflects its western location of Loveland, Colorado. The benefits of this type of design are several. Hay and equipment can be loaded into the loft from the uphill side, and the natural insulation of the hillside shelters against harsh weather. The downhill drainage away from the barn is excellent.

Built for the George Kindt family of Loveland, Colorado, the center-aisle barn is used for the family's pleasure horses, offering a shelter that is rustic in style and strong enough to withstand extreme weather and daily wear of five horses. Despite its rough-and-ready appearance, the same model can be built with more finished materials and painted if desired for a more polished, show-barn presentation.

The builder, Steve Gardner of Barns By Gardner of Berthoud, Colorado, notes that the building was constructed with rough-cut local 2" x 8" pine lumber on the siding and partitions, for both appearance and solidity. The easier air movement between the planks helps keep the horses acclimated to the outside temperatures while offering essential shelter. Applying inexpensive plywood to the inner walls would add to the weather-tight capacity of the building, without losing its aesthetic, rustic appeal. Built into the hill, the barn is generally warmer in the winter and cooler in the summer than a fully exposed building, benefiting from the hillside's protection and the heavy lumber's natural insulation qualities.

Note from the photographs that the hillside above the barn is graded to avoid runoff directly onto the structure. Such planning is essential to facilities built into the terrain this way. The flat driveway on the upper hillside also offers easy loading into the storage loft.

The interior is simple to rearrange—as with most pole barns—with the addition of a pole here or there on which to attach wall or partition sections. Additional windows or doors are also easy to add, since the walls are not supporting the roof's weight.

MATERIALS

Dimensions:
  36' x 36' x 20' 3"

Poles:
  6" x 6" x ITCCA treated (12) in 10"-15" diameter concrete-filled holes

  4" x 4" x ITCCA treated stall-door posts (10)

Concrete Revetted Wall:
  12 yards concrete for east side and partially stepped down north and south sides:

  40 lineal feet, 8' x 8" on 8" x 16" footing

  6 lineal feet, 6' x 8" on 8" x 16" footing

  6 lineal feet, 4' x 8" on 8" x 16" footing

  ½" anchor bolts, 3" o.c.

Framing:
  2" x 4" x 16' sill boards over concrete wall (4)

  2" x 12" x 12' Douglas fir, #2 or better, header beams (60)

  2" x 8" x 14' Douglas fir floor joists, 40 lbs. sq. ft., 16" o.c.

Siding:
  2" x 8" x ITCCA or redwood planks, base of siding and partitions (16)

  2" x 8" x 12' rough pine upper siding and partition planks (22)

  2" x 8" x 16' rough pine siding (20)

  2" x 8" x 10' rough pine siding (30) plus vertical supports and trim

  2" x 8" x 14' rough pine siding (10) plus vertical supports and trim on doors

Flooring:
  Hard-packed native soil, barn aisle

  ¼" CDX plywood or tongue-in-groove loft flooring,

42 sheets, with hay-drop doors cut through over each stall.

Roofing:
  2" x 6" x 16' Douglas fir rafters (54)

  2" x 6" x 14' Douglas fir rafters (28)

  54 rafter ties

  7 tubes construction adhesive

  ½" OSB, CDX plywood or ⁷/₁₆" waferboard roof sheathing

  5 rolls #15 asphalt felt

  17 sq. ft. organic asphalt 3-tab roof shingles

Exterior doors:
  1" x 8" x 12" rough pine track door planking

  Four 12' sections of barn door track, 4 trolleys, 4 rollers, 14 track hangers

Interior doors:
  20 pairs 8" strap hinges for stall doors

  20 #1135 National stall door latches

  Fifteen 4" hooks to hold hinged doors open

Interior:
  5 12' x 12' stalls, 9'6" ceilings, two sets of Dutch doors to each stall, partitions to 4' with 4" o.c. metal bars set vertically above.

  One 12' x 12' tack/feed room in center location

  12' aisle with packed dirt floor for grooming

Trim etc.:
  1" x 8" x 16' fascia boards

  2" x 6" x 16' gable-end fascias

  2" x 4" x 16' Hemlock fir framing over concrete, door stops, loft ladder, etc.

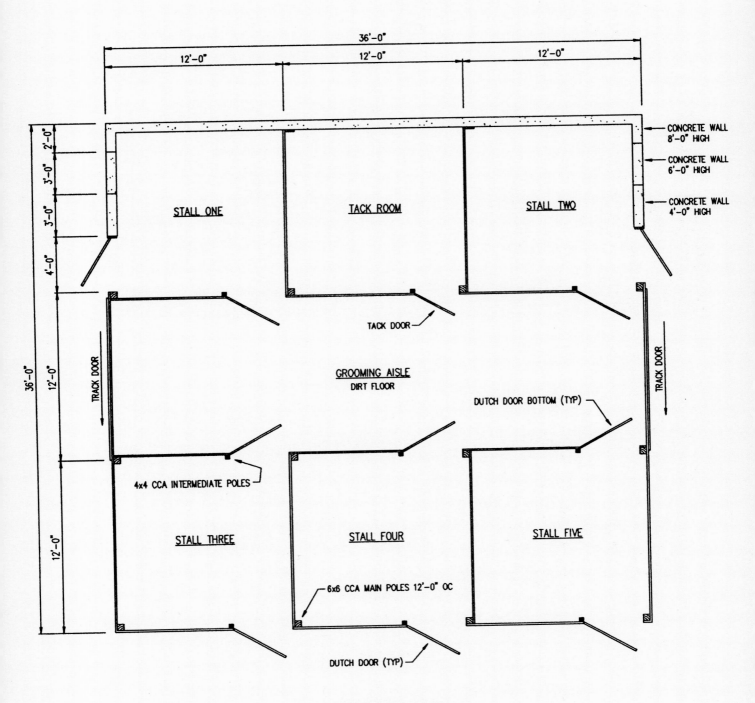

36'-0"

12'-0"   12'-0"   12'-0"

CONCRETE WALL
8'-0" HIGH

CONCRETE WALL
6'-0" HIGH

CONCRETE WALL
4'-0" HIGH

2'-0"
3'-0"
3'-0"
4'-0"

36'-0"

12'-0"

12'-0"

STALL ONE

TACK ROOM

STALL TWO

TACK DOOR

TRACK DOOR

TRACK DOOR

GROOMING AISLE
DIRT FLOOR

DUTCH DOOR BOTTOM (TYP)

4x4 CCA INTERMEDIATE POLES

STALL THREE

STALL FOUR

STALL FIVE

6x6 CCA MAIN POLES 12'-0" OC

DUTCH DOOR (TYP)

1st FLOOR PLAN

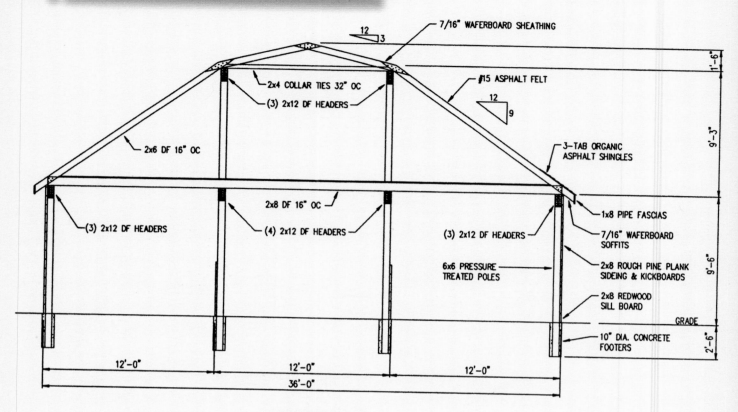

12
3

7/16" WAFERBOARD SHEATHING

2x4 COLLAR TIES 32" OC

(3) 2x12 DF HEADERS

#15 ASPHALT FELT

12
9

2x6 DF 16" OC

3-TAB ORGANIC ASPHALT SHINGLES

2x8 DF 16" OC

(3) 2x12 DF HEADERS

(4) 2x12 DF HEADERS

(3) 2x12 DF HEADERS

6x6 PRESSURE TREATED POLES

1x8 PIPE FASCIAS

7/16" WAFERBOARD SOFFITS

2x8 ROUGH PINE PLANK SIDEING & KICKBOARDS

2x8 REDWOOD SILL BOARD

GRADE

10" DIA. CONCRETE FOOTERS

1'-6"

9'-3"

9'-6"

2'-6"

12'-0"

12'-0"

12'-0"

36'-0"

SECTION

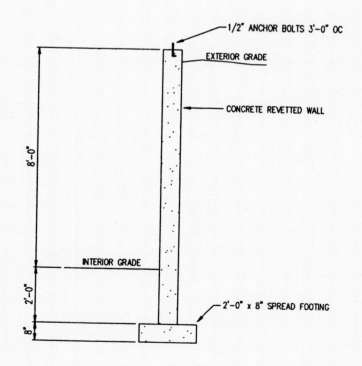

1/2" ANCHOR BOLTS 3'-0" OC

EXTERIOR GRADE

CONCRETE REVETTED WALL

8'-0"

INTERIOR GRADE

2'-0"

8"

2'-0" x 8" SPREAD FOOTING

REVETTED WALL SECTION

COLORADO BANK BARN

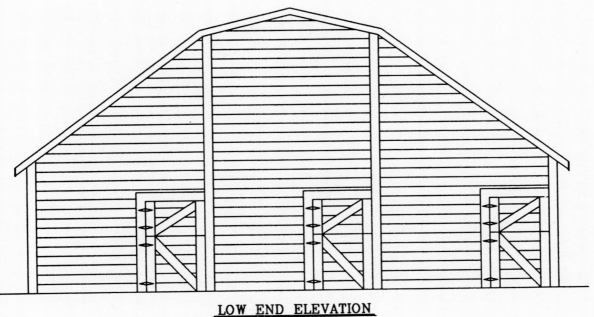

<u>LOW END ELEVATION</u>

<u>HIGH END ELEVATION</u>

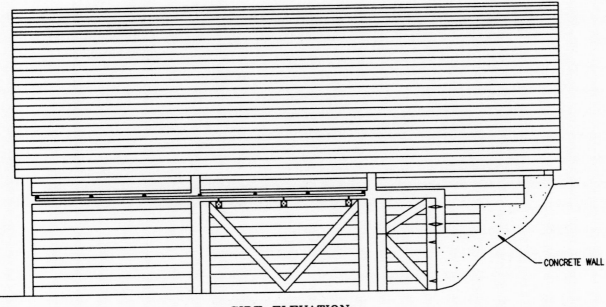

CONCRETE WALL

<u>SIDE ELEVATION</u>

The Connemara barn at Foxchase Farm in Middleburg, Virginia, was built by Upperville Barns. Foxchase is owned by Maureen Hanley. Photo by N. W. Ambrosiano

SIX-HORSE SHOW BARN

Description

A handsome barn like this is moderate in size but has all the details that make preparation for showing a pleasure: a wash stall, kitchenette, etc. The stall walls are all solid up to 4', then iron rods extend up the rest of the way, giving delightful ventilation year-round, especially when windows and doors are open. In addition, the undersides of the aisle loft were finished with a plywood ceiling, adding elegance and no open surfaces for cobwebs.

Finish it inside as elegantly as you please, or rough in the extras and just enjoy the basic, open design work areas. The materials list illustrates the barn's basic requirements; it does not reflect potential upgrades of kitchenette, bathroom, aisle ceiling, etc. This barn was designed and built by the Upperville Barn Co. of Upperville, Virginia.

MATERIALS

Poles:
  20' class 6 (6" diameter)

Framing:
  Sill girt 2" x 12"

  Siding girts 2" x 6" set 24" o.c.

  Rafter plates double 2" x 12"

  Rafters 2" x 12" set 24" o.c. with 2" x 12" ridge plate

  Collar tie girts 2" x 8"

  Collar ties 2" x 8" set 48" o.c. with ply gusset

Roof:
  ½" CDX ply

  15 lb. roofing felt

  235 1b. fiberglass shingles

Siding:
  T1-11

Trim:
  1" x 8" fascia

  l" x 6" soffit boards

  1" x 4" corner boards and casing

Doors :
  Dutch or sliding as desired on stall doors

  Aisle doors double sliders

Windows :
  Closeable shutters only

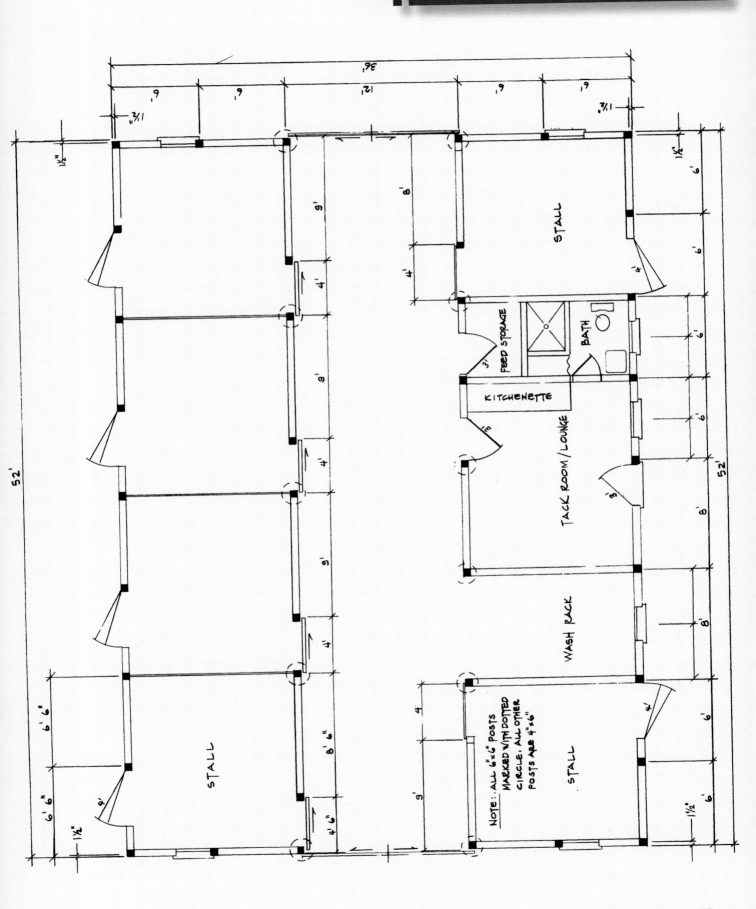

NOTE: ALL 6"x6" POSTS MARKED WITH DOTTED CIRCLE. ALL OTHER POSTS ARE 4"x6"

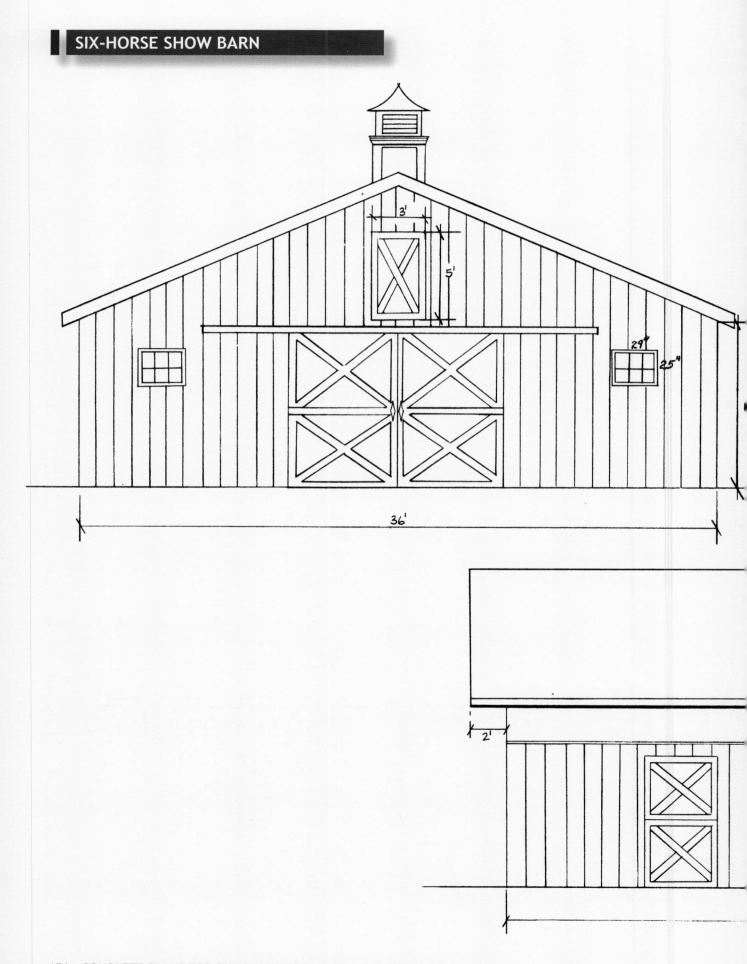

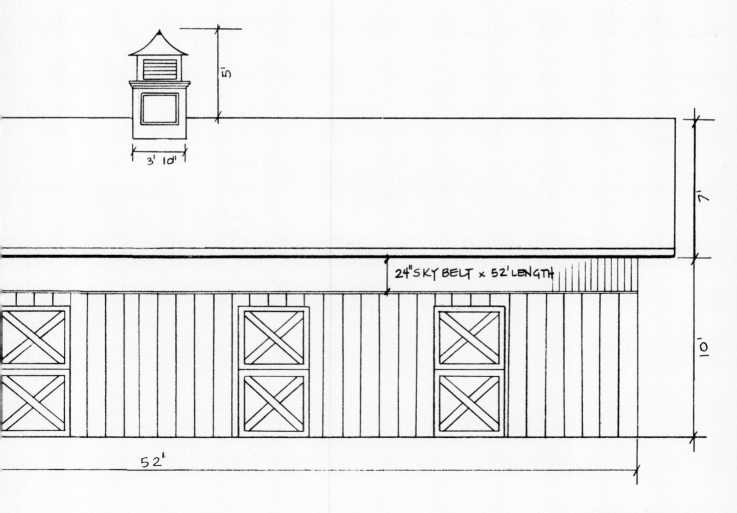

5'

3' 10'

7'

24" SKY BELT x 52' LENGTH

10'

52'

EIGHT- OR NINE-HORSE BREEDING/TRAINING BARN

Description

This barn, connected at one end to a spacious indoor arena, provides room for a busy group of horses in training as well as for their grooms. Spaces are al-lotted for wash stalls, a tack room, feed room and hay storage (only over the center aisle, providing good ventilation over the stalls). This design is from the barn building company of P. J. Williams, Inc., in Somerset, Virginia.

MATERIALS

Dimensions:
  36' x 86' (apartment 12' x 24';
  over-aisle loft 12' x 86)

Poles:
  6" x 6" down the aisle

  4" x 6" around outside of building

Framing:
  Ridge plate 2" x 12"

  Rafter plates paired 2" x 10"

  Loft floor plates paired 2" x 12"

  Loft joists 2" x 10" 16" o.c.

  Rafters 2" x 8" 4' o.c.

  Sill girts/skirt boards 3 courses 2"
  x 8" tongue-in-groove pressure-
  treated lumber

Roofing:
  ½" CDX plywood

  15 lbs. felt

  215 lbs. shingles

  2" x 4" purlins 2' o.c.

Siding:
  T1-11 textured ply

Exterior doors:
  Two pair 6' x 8' sliders for aisle
  ends, one pair 4' x 5' loft doors

Windows :
  Eight 3' x 4' single-hung
  aluminum with bars on inside set
  3" o.c.

Trim:
  Gutters on eaves with downspouts

Interior:
  Eight 12' x 12' stalls, one 12' x 14'
  foaling stall, one 12' x 12' wash
  area, one 12' x 12' tack room,
  one 12' x 14' feed room, one 12'
  x 86' over aisle loft, one 12' x 24'

apartment area, plus 6' overhang,
both eaves

Stalls :
  1" oak walls to 8' on exterior
  walls, ½" oak to 4-1/2' dividers
  with vertical bars to 8'. Yellow
  pine 2" x 6" tongue-in-groove to
  4½" stall fronts with bars to 8'.
  Slider doors to aisle 53" x 96", 4'
  x 8' Dutch door to exterior

Tack room:
  Reinforced concrete floor, ½"
  insulated ply walls, ½" lauan
  ply ceiling, 3' x 6' 8" solid core
  lauan door with locking handle.
  Insulated aluminum single-hung
  window. Laundry tub and washer
  and dryer hookups installed

Wash room:
  Reinforced concrete floor with
  brushed finish, slope to drain; 4"
  concrete block walls with stucco
  finish; one 4' x 8' Dutch door to
  exterior

Apartment:
  Reinforced concrete floor, walls
  paneled with 6" insulation on
  exterior, fireproof sheetrock
  on walls adjacent to stable;
  ¼" lauan ply ceiling with 6"
  insulation above; 3' x 4' single-
  hung aluminum windows and
  screens; hollow-core lauan
  interior doors; exterior doors to
  aisle and outside 3' x 6' 8" steel
  x-buck with locking handles.

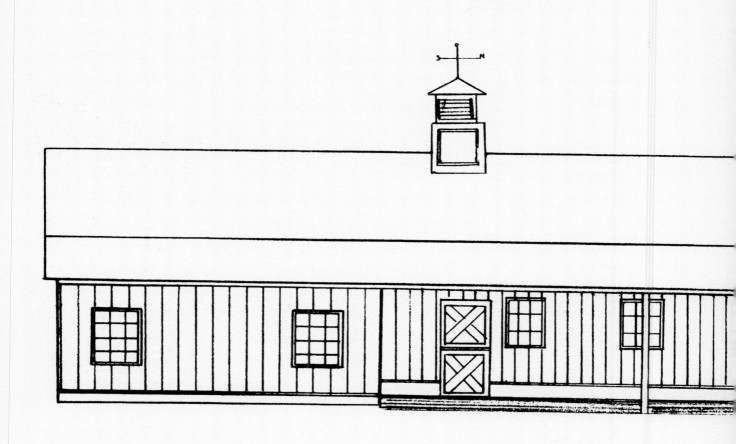

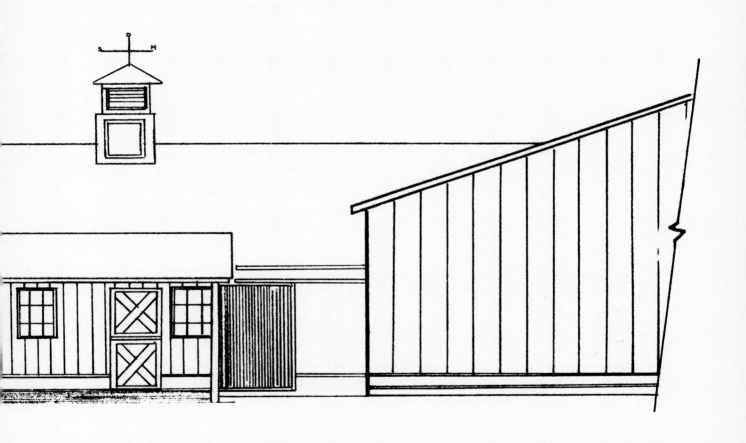

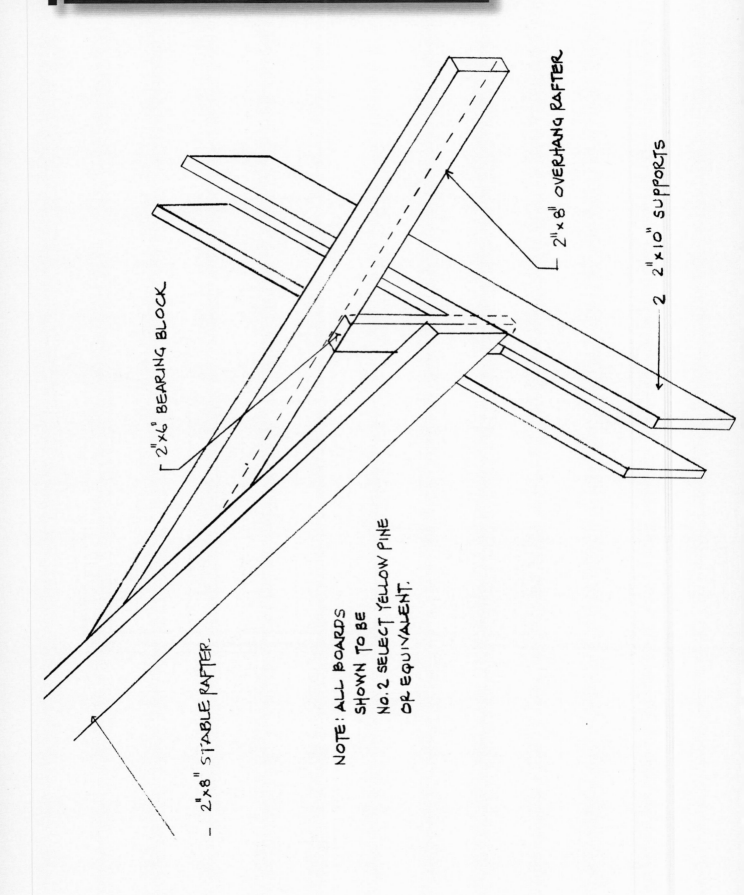

2"×8" OVERHANG RAFTER

2 2"×10" SUPPORTS

2"×6" BEARING BLOCK

NOTE: ALL BOARDS
SHOWN TO BE
No.2 SELECT YELLOW PINE
OR EQUIVALENT.

— 2"×8" STABLE RAFTER.

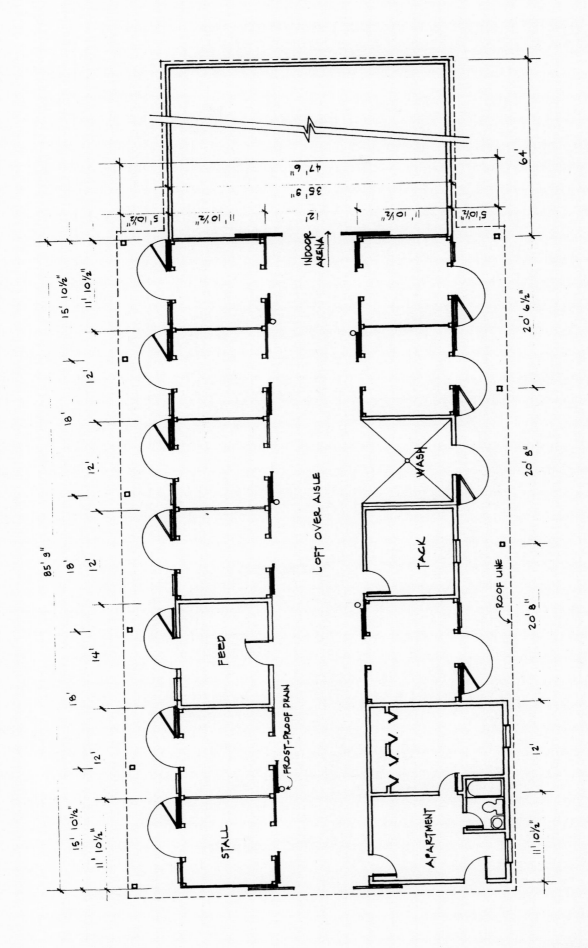

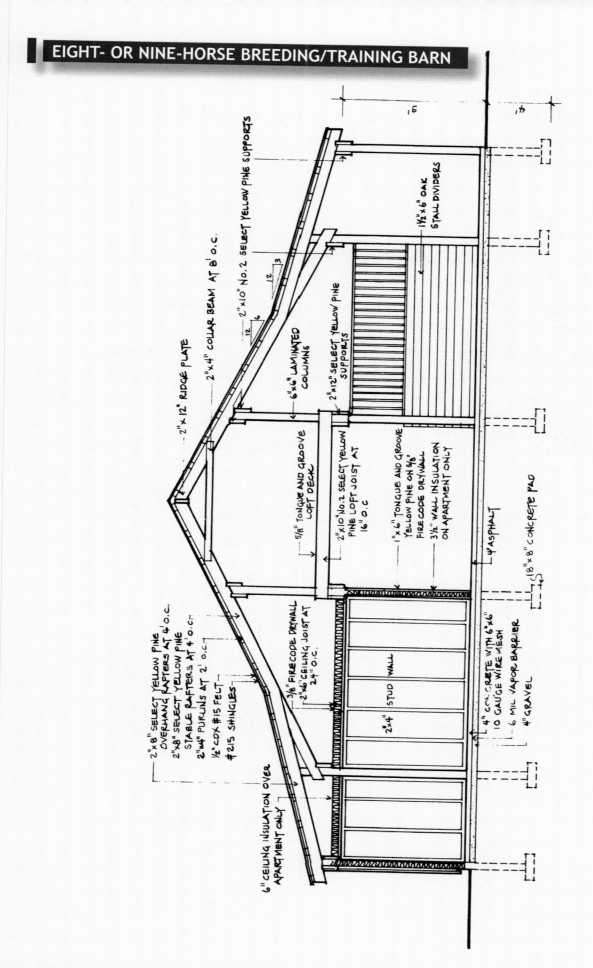

FOUR- OR SIX-HORSE BARN WITH COTTAGE ATTACHED

Description

With this attached barn and cottage, you'll no longer have to worry about a long drive to the barn or getting cold on the way. The cottage, as shown in the plans, is on a poured slab with a cinder block foundation, but it can easily be built using the same pole construction techniques as used for the barn. Hay storage is over the center aisle, and tack, storage and feed areas are split off from the regular stall spaces. Be sure the firewall arrangements between your stable and living areas conform to your local codes and that they offer the level of protection you want. This barn was built by Upperville Barns for Joe Ann Scott in Aldie, Virginia.

MATERIALS

Poles:
   20' Class 6 (6" diameter) poles

Framing:
   Sill girts 2" x 12"

   Siding girts 2" x 4" set 20" o.c.

   Rafter plates double 2" x 12"

   Rafters 2" x 12" set 24" o.c. with 2" x 12" ridge plate

   Collar tie girts 2" x 8"

   Collar ties 2" x 8" set 48" o.c. with ply gusset

   In house only, interior 2" x 4" girts with insulation and fire-resistant drywall

Roofing:
   ½" CDX ply with clips

   15 lbs. roofing felt

   235 1bs. fiberglass shingles

Siding:
   T1-11

Trim:
   1" x 8" fascia

   1" x 6" soffit boards (split for ventilation)

   1" x 4" corner boards and casings

Doors:
   2" x 6" stiles and ¾" ply backing

   2" x 6" cross buckles

Windows:
   Standard 2-over-4 window in office; others, barn sash sliding

Upperville Barns built this handsome cottage/barn combination for Joe Ann Scott in Aldie, Virginia. The barn has room for a maximum of six horses, hay storage over the center aisle only, and a small attached cottage/apartment. Fiberglass panels below the eaves improve the lighting, and plenty of customization inside provides a barn that exactly fits Scott's requirements. Photo by M. F. Harcourt.

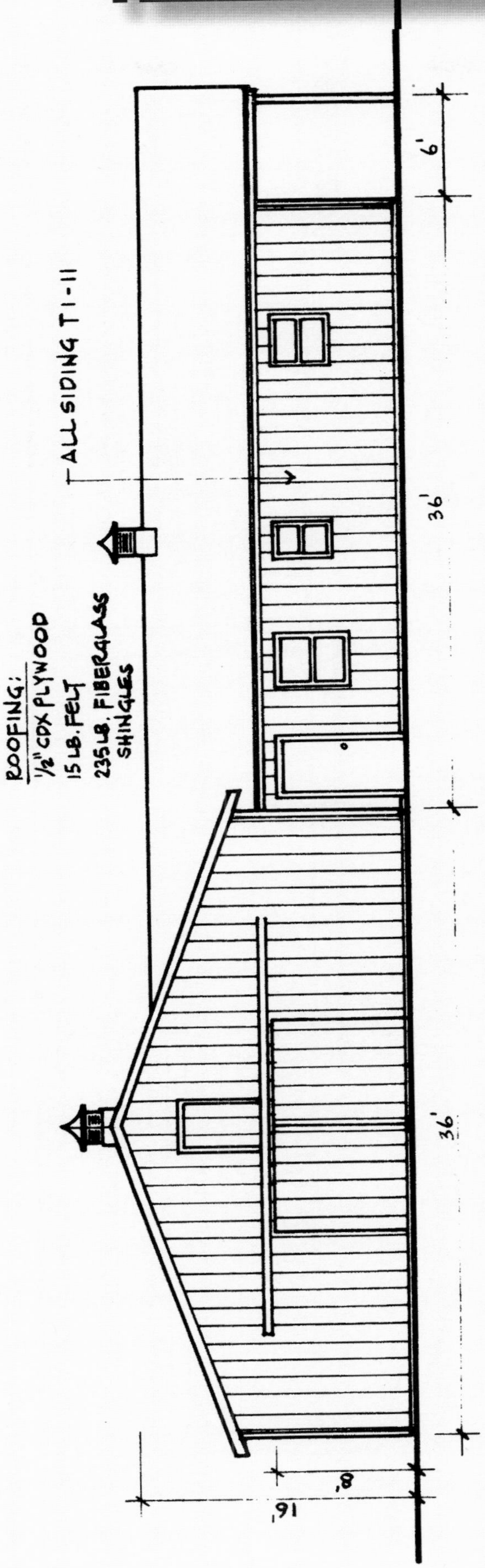

FOUR- OR SIX-HORSE BARN WITH COTTAGE ATTACHED

ALL SIDING TI-II

ROOFING:
½" CDX PLYWOOD
15 LB. FELT
235 LB. FIBERGLASS
SHINGLES

6'

36'

36'

8'

19'

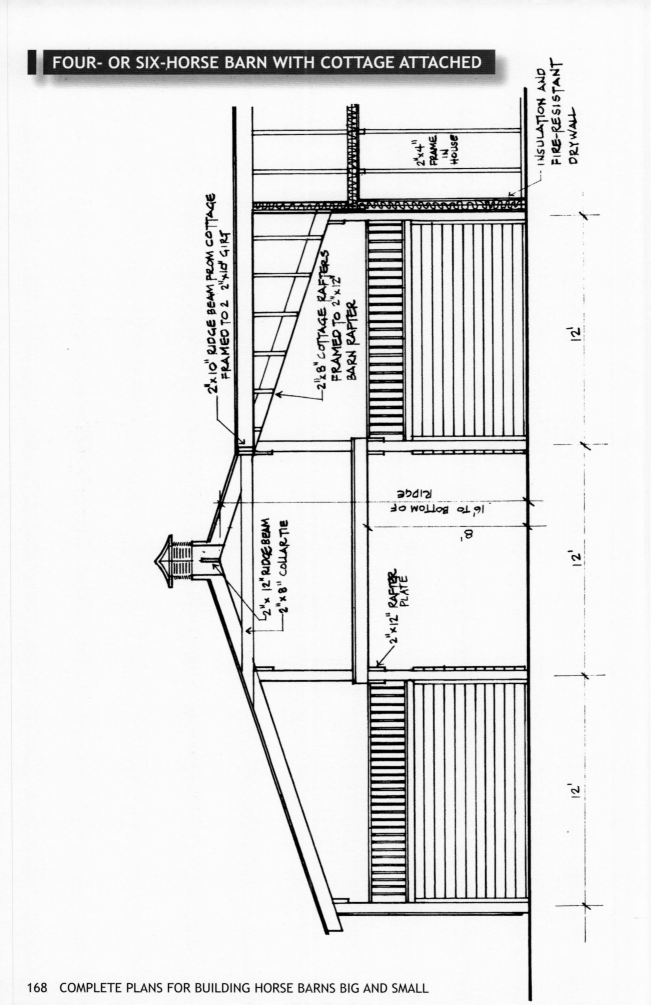

INSULATION AND FIRE-RESISTANT DRYWALL

2"x4" FRAME IN HOUSE

2"x10" RIDGE BEAM FROM COTTAGE FRAMED TO 2 2"x10" GIRT

2"x8" COTTAGE RAFTERS FRAMED TO 2"x12" BARN RAFTER

2"x12" RIDGE BEAM
2"x8" COLLAR TIE

16 TO BOTTOM OF RIDGE

8'

2"x12" RAFTER PLATE

12'

12'

12'

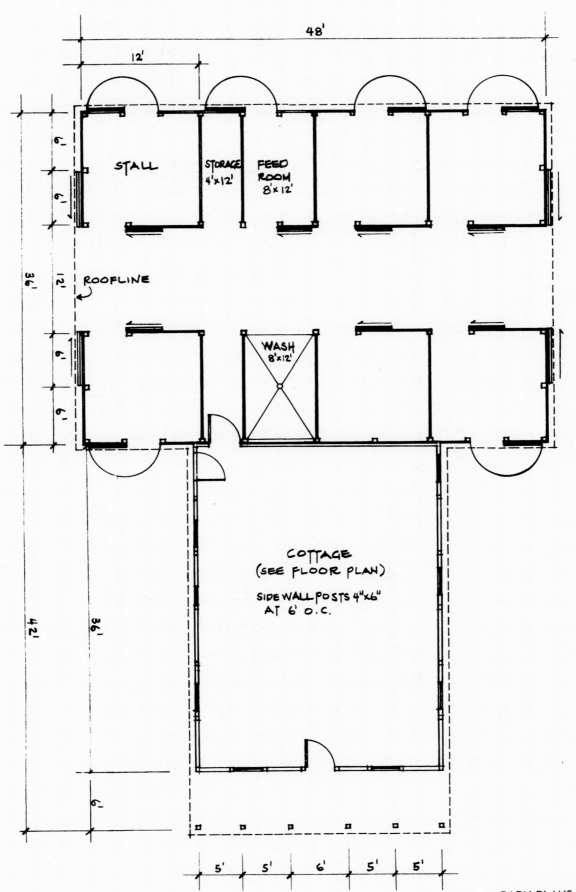

48'

12'

6'

6'

STALL

STORAGE
4'x12'

FEED
ROOM
8'x12'

ROOFLINE

36'

12'

6'

6'

WASH
8'x12'

36'

42'

COTTAGE
(SEE FLOOR PLAN)

SIDE WALL POSTS 4"x6"
AT 6' O.C.

6'

5'   5'   6'   5'   5'

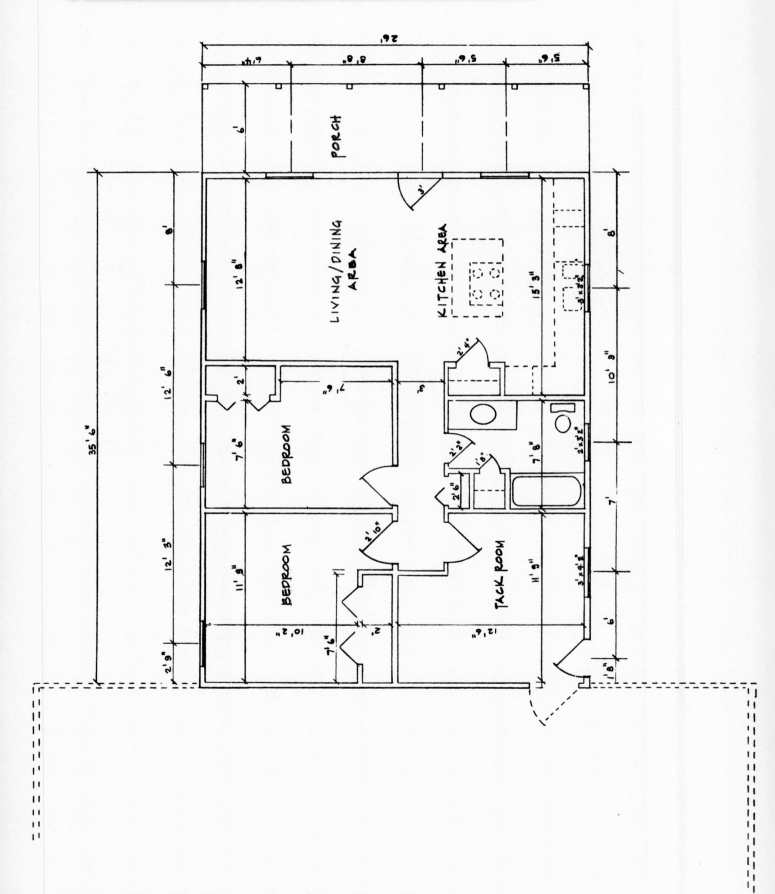

EIGHT-HORSE BARN WITH HOME OVERHEAD

Description

This handsome home and barn combination is the design of Virginia architect David Cooper and his wife, Sharon, who received advice from the Umbaugh Pole Building Company. It combines standard pole barn construction with a little of what its owners call *frou-frou,* such as the louvered vents at either end of the aisle doors. The vents fill the space added when the 10' aisle was expanded to 14' so that the 2,400-square foot upstairs home would not have obtrusively placed columns.

The tack room doubles as furnace room and entrance to the upstairs home. To bring light and style to this high-traffic area, the Coopers decided to use glass walls. In fact, glass is the essential ingredient that lends much of the home's charm, with balconies cut through the roof in three directions.

Inside the barn, the Coopers used tongue-in-groove lumber to provide a polished, neat look. A layer of fire re-

tardant drywall along the ceilings keeps the insulation, household wiring and plumbing from the horses' reach.

In 2003, this barn took direct hits from both a hurricane and a tornado. The only damage was to the roof from the tornado, requiring the roofing to be replaced. The sturdy pole structure held up with no damage.

If you decide to build this unique house and barn combination, make sure it meets your local codes, the BOCA requirements for your area, and your insurance company's regulations.

The southern exposure of the building allows maximum sunshine into the upstairs home. Stalls, at left, are shaded from both sun and rain. (Cooper Farm, Aldie, Virginia) Photo by N. W. Ambrosiano.

This photo shows the rear construction view of this basic pole barn. The idea was supplied by the Umbaugh Barn Company of Pennsylvania; David Cooper, the architect, blended all of Sharon's preferences into the final equation. The balcony, accessed by sliding glass doors, affords a view of the pastures where the Coopers keep their horses. It fits neatly over the 14' wide center aisle Sharon requested. (Cooper Farm, Aldie, Virginia) Photo by N. W. Ambrosiano.

MATERIALS

Poles:
  6" x 6" pressure-treated

Framing:
  Sill girts double 4" x 4" p.t. sections

  Rafters 12" x 12" or Grade 2 10" x 2" at 24" o.c. coupled at posts with ½" CDX sheathing ply, joined at midpoint of column with bolted connectors

  Purlins spaced flush with 10" x 2" column post

  Joists 10" x 2" 24" o.c. with purlins finished with 5/8" drywall fire screen on underside

  Interior stable framing 4" x 4"

  Stalls lined with 2" x 6" tongue-in-groove oak to 42" height, metal rods set 3" o.c. vertically

  Deck boards 2" x 4" on bearers above galvanized pan

  Upstairs 2" x 6" stud wall framing 2' 0" o.c.

Exterior:
  T1-11 with cedar trim, louvers at either side of exterior sliders

Flooring:
  Rough-finished concrete aisle: stalls, packed soil

Outside stall area beneath overhand 12" x 12" pavers edged in brick

Upstairs ¾" ply under-layment fixed and glued, kitchen/bath/laundry tile floor, hardwood and carpets in bedrooms, living room, etc.

Doors :

2 sliders either end

Interior sliders each stall, with central window filled by 3" o.c metal rods

Dutch doors to exterior framed with 2" x 6", faced with same ply sheathing as wall

Exterior entrance to tack room and stairs area walled with fixed glass screen and fully

glazed door

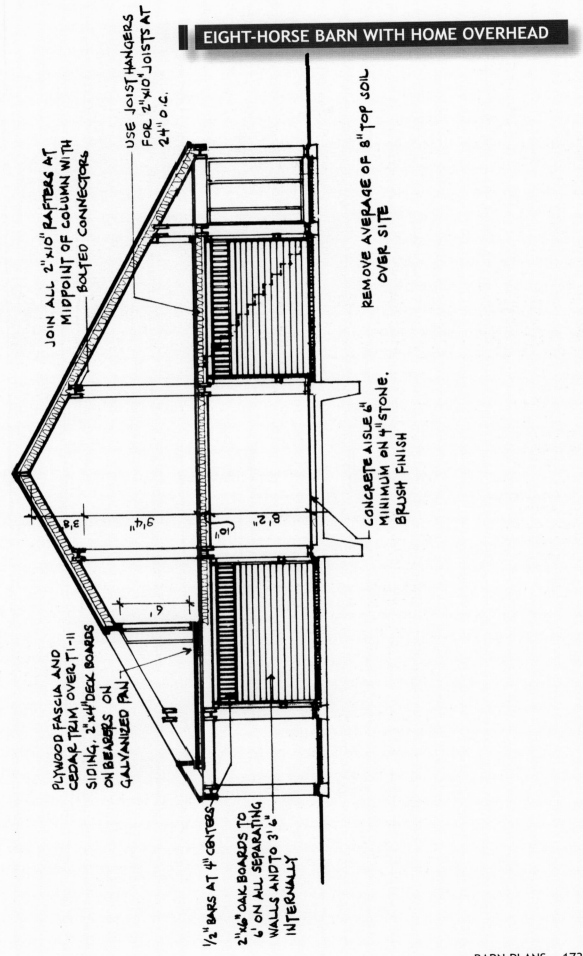

USE JOIST HANGERS FOR 2"x10" JOISTS AT 24" O.C.

JOIN ALL 2"x10" RAFTERS AT MIDPOINT OF COLUMN WITH BOLTED CONNECTORS

REMOVE AVERAGE OF 8" TOP SOIL OVER SITE

CONCRETE AISLE 6" MINIMUM ON 4" STONE. BRUSH FINISH

3'8"

3'4"

10'

8'0"

6'

PLYWOOD FASCIA AND CEDAR TRIM OVER T1-11 SIDING, 2"x4" DECK BOARDS ON BEARERS ON GALVANIZED PAN

1/2" BARS AT 4" CENTERS

2"x6" OAK BOARDS TO 6' ON ALL SEPARATING WALLS AND TO 3'6" INTERNALLY

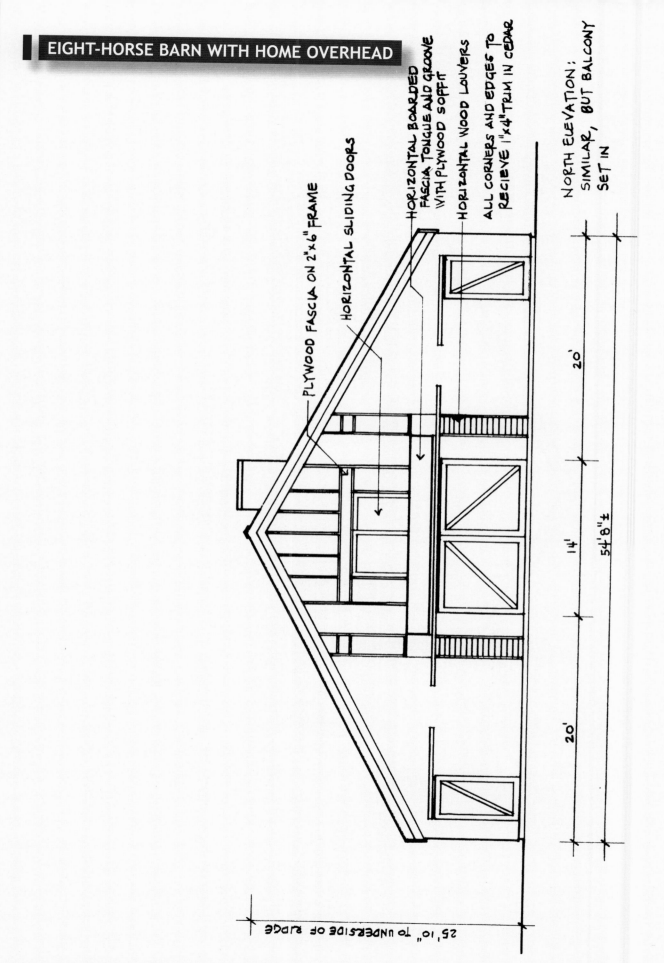

PLYWOOD FASCIA ON 2"x6" FRAME

HORIZONTAL SLIDING DOORS

HORIZONTAL BOARDED FASCIA TONGUE AND GROOVE WITH PLYWOOD SOFFIT

HORIZONTAL WOOD LOUVERS

ALL CORNERS AND EDGES TO RECIEVE 1"x4"TRIM IN CEDAR

NORTH ELEVATION; SIMILAR, BUT BALCONY SET IN

25'10" TO UNDERSIDE OF RIDGE

20'

14'

54'8"±

20'

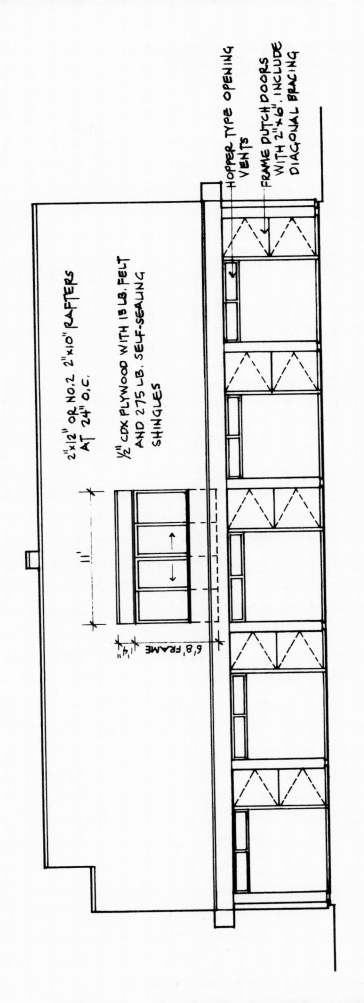

HOPPER TYPE OPENING
VENTS

FRAME DUTCH DOORS
WITH 2"x6". INCLUDE
DIAGONAL BRACING

2"x12" OR NO.2 2"x10" RAFTERS
AT 24" O.C.

½" CDX PLYWOOD WITH 15 LB. FELT
AND 275 LB. SELF-SEALING
SHINGLES

11'

6'8' FRAME
1'4"

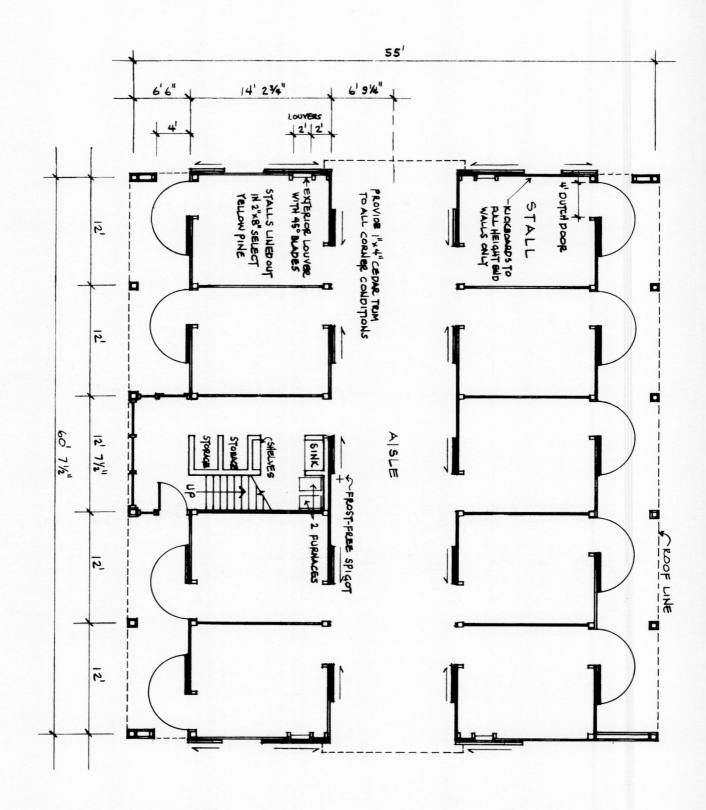

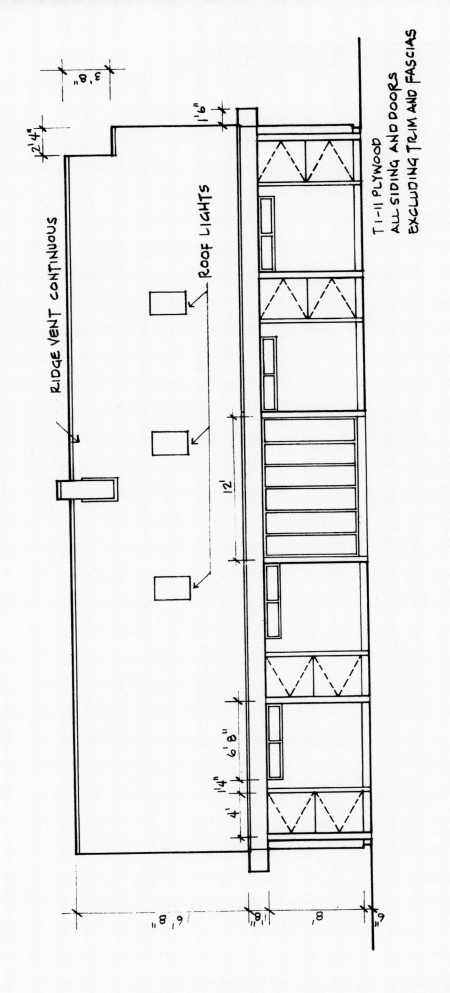

RIDGE VENT CONTINUOUS

ROOF LIGHTS

T 1-11 PLYWOOD
ALL SIDING AND DOORS
EXCLUDING TRIM AND FASCIAS

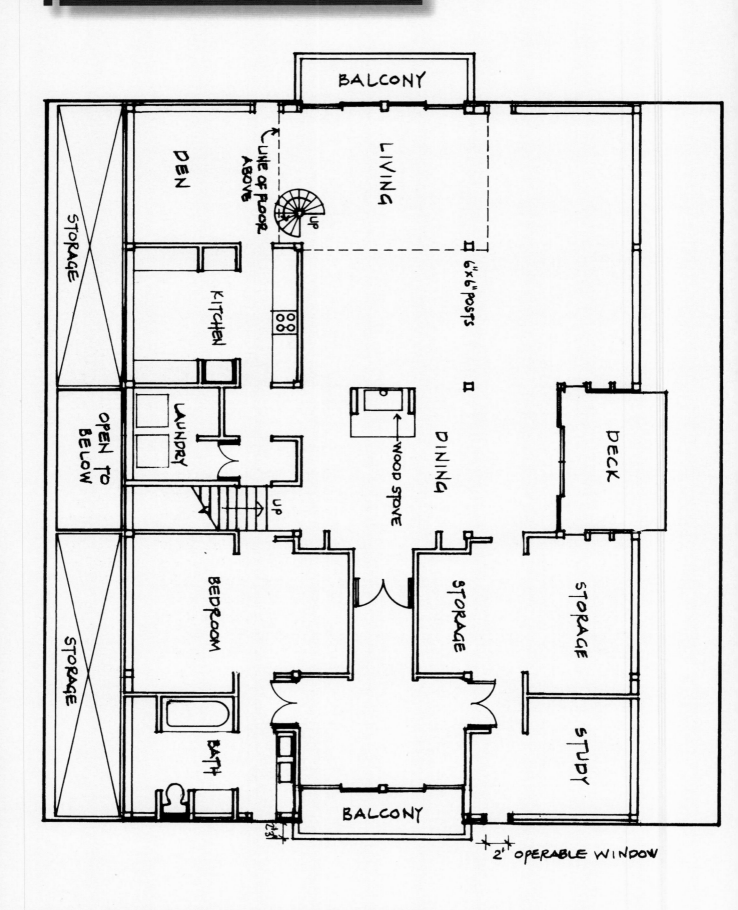

BALCONY

DEN

LIVING

LINE OF FLOOR ABOVE

UP

STORAGE

KITCHEN

6"x6" POSTS

OPEN TO BELOW

LAUNDRY

DINING

WOOD STOVE

DECK

UP

BEDROOM

STORAGE

STORAGE

STORAGE

STUDY

BATH

BALCONY

2' OPERABLE WINDOW

CONVERTIBLE BARN/ GARAGE

Description

This barn is ideal for a home in suburbia that needs re-sale flexibility. The basic design and structure are built for a barn. Modifications are then made to the exterior walls to allow garage door facades placed on the street side of the barn. Stall partitions and walls are added to the interior, and can be easily removed. The same materials are used to create interior walls over the garage door openings, establishing stalls and work areas in the barn. When no longer needed, the interior stall walls can be removed. Garage doors can be added, and the once-working stable becomes a fully usable garage.

The loft is designed to hold a full 20 tons of hay, although it can be converted to a small apartment or made with a lower pitch to the roof. This barn was built on a cinderblock foundation to overcome a very wet piece of land; it can be built as a pole structure as well. It was designed and built by former owners Penny and Lloyd Burger of Chesapeake, Virginia.

This barn, originally owned and designed by Penny and Lloyd Burger of Chesapeake, Virginia, can be converted to a garage in a matter of a few hours. Built in an area of difficult drainage, it sits on a cinderblock foundation. It is faced with T1-11 textured-plywood stained to match the nearby house; the asphalt-shingle roof also matches the house. Photo by N. W. Ambrosiano

Eight-foot garage-door openings in the walls of this convertible barn face toward the house. Each was filled with 2" x 8" planks topped with metal bars for ventilation and light. (Burger Farm, Chesapeake, Virginia) Photo by M.F. Harcourt.

MATERIALS

Dimensions:
  30' x 36'

Poles :
  6" x 6" inside supports on concrete footers, outer framing 4" x 4" and 2" x 4" platform style frame on cinderblock foundation

Framing:
  Loft floor joists 2" x 12" set 16" o.c. supported by either built-up wood beam of 2" x 12" or steel I-beam on 6" x 6" post with concrete footers. Garage door openings framed with two 2" x 4" jacks to support two 2" x 10" headers

Roofing:
  ½" plywood deck with roofing felt and asphalt shingles to match nearby structures

Siding:
  1" x 12" rough-cut barn siding, or T1-11 2" x 8" x 8' rough-cut oak in doorways, topped with vertical rods 3" o.c.

Trim:
  1" x 6" starter boards at corners, 1" x 6" trim boards in garage openings, 1" x 6" fascia boards

Doors:
  Two 5' sliders at each end, six 4' x 8' sliders for stalls, or screens as desired

Windows :
  3' x 3' openings framed with 2" x 4", with rods set 3" o.c.

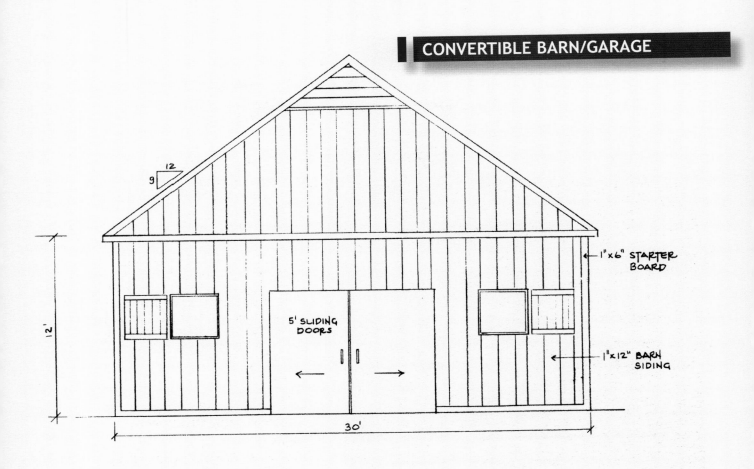

12
9

1"x6" STARTER
BOARD

12'

5' SLIDING
DOORS

1"x12" BARN
SIDING

30'

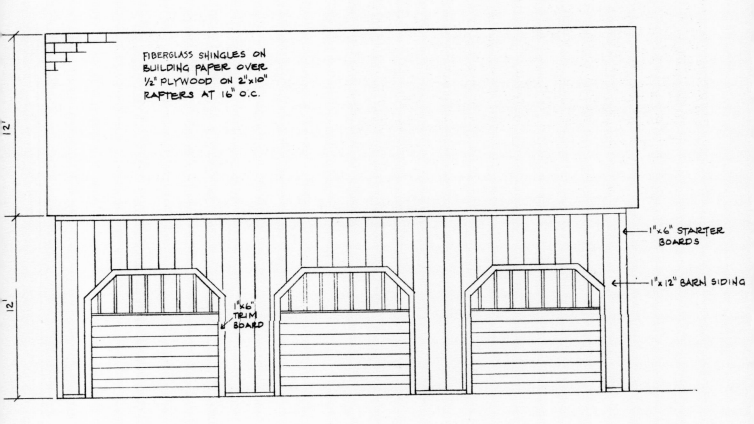

FIBERGLASS SHINGLES ON
BUILDING PAPER OVER
½" PLYWOOD ON 2"x10"
RAFTERS AT 16" O.C.

12'

1"x6" STARTER
BOARDS

1"x12" BARN SIDING

12'

1"x6"
TRIM
BOARD

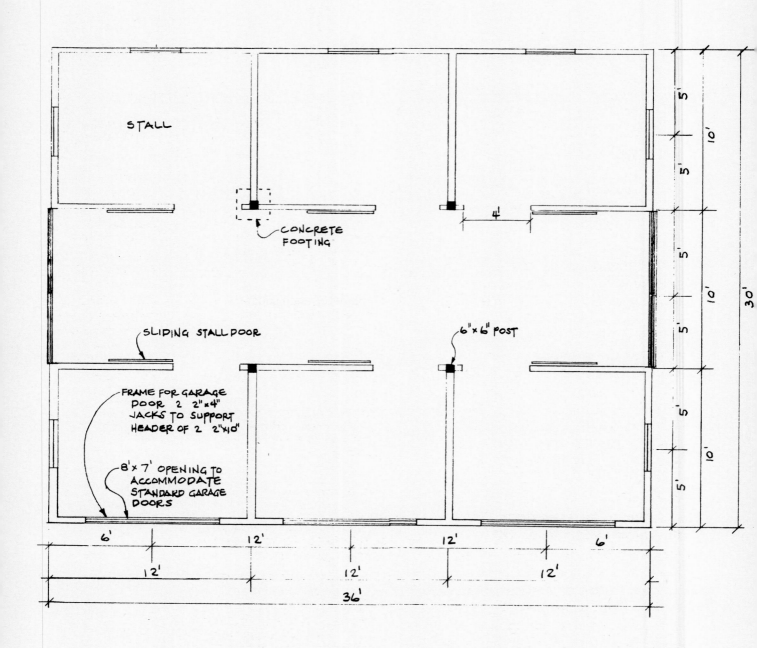

STALL

CONCRETE FOOTING

4'

SLIDING STALL DOOR

6"x 6" POST

FRAME FOR GARAGE DOOR 2 2"x4" JACKS TO SUPPORT HEADER OF 2 2"x10"

8'x 7' OPENING TO ACCOMMODATE STANDARD GARAGE DOORS

5' 5' 10' 5' 5' 10' 5' 5' 10' 5' 30'

6' 12' 12' 6'
12' 12' 12' 12'
36'

HOT-WEATHER BARNS

Description

The numerous old and new facilities located throughout the warm states, while reflecting varied climates and needs, have several features in common.

Larger aisleways or smaller stall units in each structure either remove excess heat or reduce the amount of heat by having smaller numbers of horses per area.

They contain larger-than-normal ventilation areas. Large windows, full-length roof vents/cupolas, half-solid stall partitions, or wide gaps at the tops of full partitions to allow air circulation are some of the construction features that encourage ventilation. Often doors are full-length, heavy-gauge wire to allow increased circulation.

These barns are strategically located in relation to the prevailing winds and terrain, taking full advantage of both. They use natural shade from trees whenever possible.

The simple pole barn on page 187 represents the most versatile of buildings, actually being a wall-less collection of stalls beneath an airy roof. While shown with heavy lumber here, as might suit a draft-horse facility, the stalls can be framed in rough-cut 1" x 6"s with 3"

MATERIALS

(for pole barn)

Dimensions:
  40' x 240' (modifiable in all directions)

Poles :
  8" diameter main supports, treated poles or telephone poles

Framing:
  Rafter girts 2" x 8"
  Rafters 2" x 8" with purlins 24" o.c.

  Stall posts 4" x 4" posts or 4" diameter poles with rough-cut 2" x 6" or larger partitions spaced 3"

Roof:
  Corrugated metal

Details:
  Tack rooms and feed areas can be substituted for any stalls by using ½" ply walls over

  2" x 4" girts instead of rough-cut board partitions. Aisle and stall dimensions can be expanded as desired, since roof weight is light and supporting poles/rafters are sufficient to accept wider span.

This tile-roofed California breeding farm makes use of the long growing season, spreading cool ivy around the overhang, but containing the growth around the posts in horse-proof wire mesh. The breeze beneath is cool and soft, no matter the weather. (Nicasio Arabians, Nicasio, California) Photo by N. W. Ambrosiano.

gaps between boards for additional air flow.

Other methods of providing relief for barns in hotter climates can come from architectural designs native to

An alternate structure for hot weather climate is constructed of cinderblock. While expensive to construct, mainly due to labor costs, the cinderblock barn, as shown, can be a cool one. This structure incorporates all the best that natural ventilation can offer.

Large trees offer natural shade. In addition, the lay of the facility takes full advantage of a natural draw between the hills, channeling any breezes or natural air flow down through the center aisle.

The center aisle is exceptionally wide as are the doors and windows. In the interior, the upper area has been left totally open, with vents at each end. Doors and walls were constructed with bars on the upper half to increase airflow.

Many safety features have been incorporated into this barn, such as rounding all exposed cinderblock corners to reduce nicked horse hips

Since this is a training facility, with frequent shipments of young stock to be broken for the track, this barn is constantly being thoroughly cleaned in preparation for a new crop of highly susceptible youngsters. (September Farm, Ocala, FL)

This California barn features the mission-style tile roof, stucco walls and rounded windows typical in the housing of the southwest and west. Inside, open rafters above the stalls and aisle doors that open in four directions capture every passing breeze. (Windfield Station, Nicasio, California) Photo by N.W. Ambrosiano.

A variation on the cinderblock structure allows the stalls to be backed up to each other with the aisleway fronting them on the outside. Ventilation has been increased by adding full heavy-gauge wire doors along with wide ventilation strips on the three inside walls. Since this is a breeding facility, stall separations are full, and air flow is maintained by the addition of venting blocks spaced strategically in the walls. Full use of natural shade from the trees also keep this facility cool. (Hassell Arabian Stud, Reddick, Florida) Photo by M. F. Harcourt.

The importance of sturdy interior construction is obvious when you examine the residents in their stalls. (Briar Patch Farm, Micapony, Florida) Photo by M. F. Harcourt.

This pole barn structure is very basic yet fully functional for any of a number of horse (or livestock) operations. Designed to take advantage of the South's warm breezes, this barn forgoes outer walls and its design allows great creativity with the inner layout. Additional poles can be sunk at any location under the freestanding span to allow enclosed tack rooms, feed areas, wash stalls or trailer storage. The exceptionally high roof provides shade and allows the heat to rise and then dissipate. This farm raises Shires, and these gentle giants rarely cause problems, including the stallions. But the heavy lumber framing for the stalls can withstand almost any tough actor. The owner has added a safety feature in the double-wide front stall that holds a mare and foal. The lower area of the walls has been completely enclosed to keep a young foal from sticking a leg through and being injured. (Briar Patch Farm, Micapony, Florida) Photo by M. F. Harcourt.

This variation on the standard open stall makes use of tie stalls. This allows more animals to be handled in the same area, which is ideal for horses used constantly throughout the day. They can then stand quietly between sessions. These stalls are not recommended for a horse's full-time shelter, however, as they are far too confining. (Briar Patch Farm, Micapony, Florida) Photo by M. F. Harcourt.

The aisleways in this airy pole barn can be constructed so they are wide enough for working with any equipment or animal. The feed room is the enclosed area to the left that has been placed on a concrete floor to protect the feed. The center aisle is made of asphalt for easy cleaning and to facilitate the movement of the heavy vehicles used by the owner. (Briar Patch Farm, Micapony, Florida) Photo by M. F. Harcourt.

countries with those warmer climates. Using natural lays of the land to create air flows or natural or artificial shade helps to increase air exchange and decrease the temperature effects of bright sunlight.

Stucco or cinderblock walls, with scattered ventilation spaces, help to reduce interior temperatures. Roofs that reflect, rather than absorb, heat energy from the sun also help to maintain a cool interior. High ceilings are an advantage as they let hot air rise past the stall area upward where it can be vented out through either natural or artificial means. Remember that insulation can be used to help keep heat out as well as in.

Also, consider Spanish or Moorish architecture that incorporated vegetation and fountains or running water along with good ventilation and shad-

ing to keep heat from building up or to help it dissipate when it did.

A visit to a company with knowledge on solar energy may give you some good ideas toward keeping that hot-climate barn comfortable for man and beast.

If you are building your barn in an area where winters can be bitter, do your research on what local builders and the extension service recommends. Check on the local code requirements to get an idea of the pitfalls and problems you must overcome with such extreme weather conditions. In areas of high snowfall, note the steep pitch to older roofs that have stood the test of time to get an idea of what you may need to plan for.

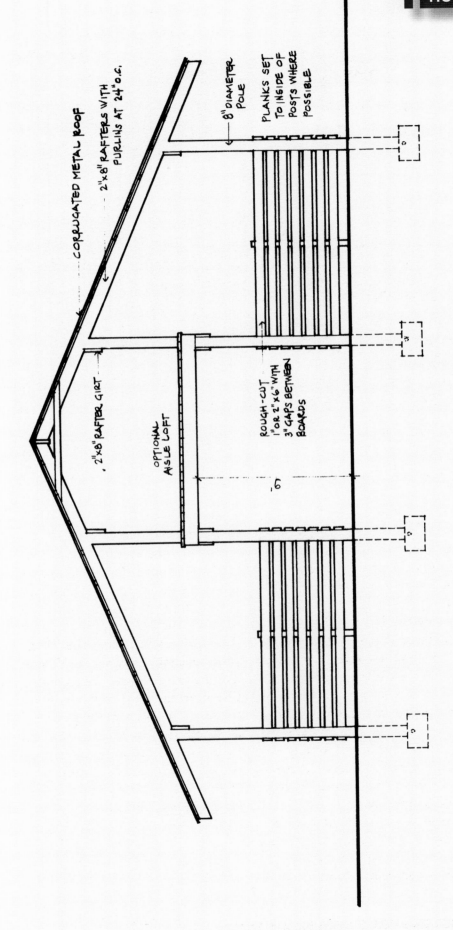

CORRUGATED METAL ROOF

2"x8" RAFTERS WITH PURLINS AT 24" O.C.

8" DIAMETER POLE

PLANKS SET TO INSIDE OF POSTS WHERE POSSIBLE

2"x8" RAFTER GIRT

OPTIONAL AISLE LOFT

ROUGH-CUT 1" OR 2"x6" WITH 3" GAPS BETWEEN BOARDS

6'

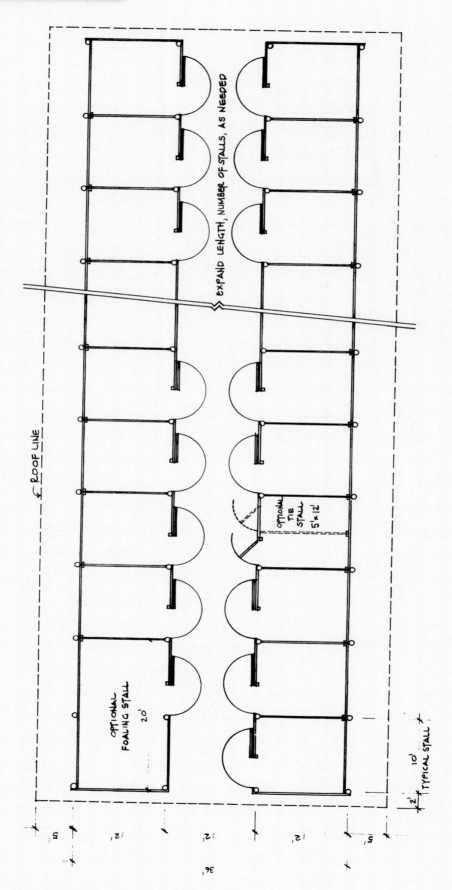

TRAILER SHED WITH RUN-IN AND HAY STORAGE

Description

This simple building was designed and built by Captain and Mrs. Douglas A. Williams while living in Chesapeake, Virginia. It began as a 20' x 40' trailer shed with a loft to hold 3½" tons of hay. On both sides, shed roofs were built to accommodate more equipment. A run-in shed was added where the building abuts a field.

Dimensions in this building are approximate, allowing for the minor changes and inaccuracies of a casual building project. The side sheds for run-in and equipment are shown with three support poles, but can easily be built with four poles, depending on your equipment's dimensions.

MATERIALS

Dimensions:
  44' x 40'

Posts:
  6" x 6" pressure-treated

Framing:
  2" x 8" paired rafter plates

  2" x 8" rafters 24" o.c.

  2" x 8" ridge plate supported by four 4" x 4" posts set on loft floor

Siding:
  T1-11 for back wall and partial sides

  2" x 8" or larger planks for run-in and side storage walls

Roofing:
  4' x 12' aluminum roofing nailed over 2" firring strip or 2' x 4' purlins

Trim:
  Gutters along rafter eaves

Door:
  Site-built loft door

This is a good example of a workable trailer, hay and run-in shed. (Captain and Mrs. D. A. Williams, Chesapeake, Virginia) Photo by N.W. Ambrosiano.

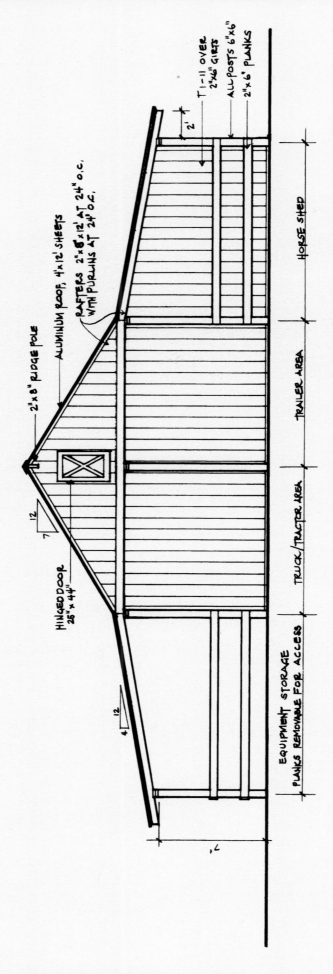

T 1-11 OVER 2"x6" GIRTS

ALL POSTS 6"x6"

2"x6" PLANKS

2'

HORSE SHED

TRAILER AREA

TRUCK / TRACTOR AREA

EQUIPMENT STORAGE
PLANKS REMOVABLE FOR ACCESS

ALUMINUM ROOF, 4'x12' SHEETS

RAFTERS 2"x6"x12' AT 24" O.C.
WITH PURLINS AT 24" O.C.

2"x 8" RIDGE POLE

12
7

HINGED DOOR
25"x 44"

12
4

2'

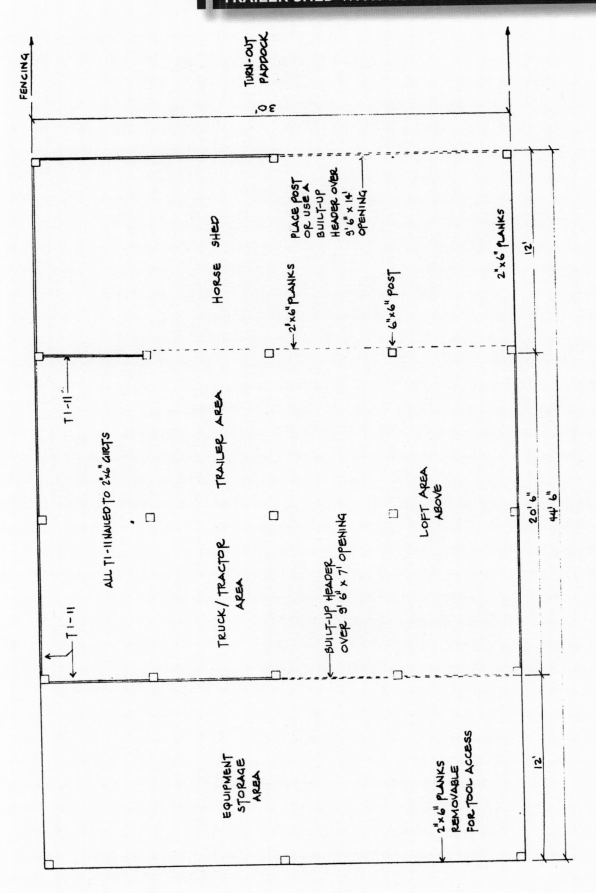

FENCING

TURN-OUT PADDOCK

30'

HORSE SHED

PLACE POST OR USE A BUILT-UP HEADER OVER 3'6" x 14' OPENING

2"x6" PLANKS

6"x6" POST

2"x6" PLANKS

12'

TI-11

ALL TI-11 NAILED TO 2"x6" GIRTS

TRAILER AREA

TI-11

TRUCK/TRACTOR AREA

BUILT-UP HEADER OVER 3' 6" x 7' OPENING

LOFT AREA ABOVE

20' 6"

44' 6"

EQUIPMENT STORAGE AREA

2"x6" PLANKS REMOVABLE FOR TOOL ACCESS

12'

Sheltered storage and covered work areas are the hallmark of this compact barn. One of several models available from Morton Buildings, this barn features a finished steel exterior and spacious stalls.

TWO-STALL MANUFACTURED BARN

Description

This small barn package is an excellent arrangement for areas that get some heavy weather. With plenty of sheltered storage and work areas in a small space, a conservative horse owner can spend plenty of time with the animals and not venture out into cold winds or rain until it's time to either ride or leave.

The design is by John Stover of the Culpeper, Virginia, office of Morton Buildings, Inc., and represents but one of a range of plans offered. As with other top quality barn companies, regional offices can work with horse owners for a plan based on the specific requirements of their land, their horses and their climate. This company has branch offices in 35 states and a home office in Morton, Illinois, so regions of the country unfamiliar with them are few.

This barn, built according to Culpeper County requirements, prices out at approximately $34 per square foot if placed on a prepared site. This will vary dramatically according to your area's building codes, climatic change and your choice of materials.

Most of the building materials are manufactured by Morton's own factories and cut for individual plans. This barn is shown with a finished steel exterior and steel wainscoting, a choice offered by professional barn companies that isn't as easily available to independent owner/builders.

MATERIALS

Dimensions:
   24' x 36'

Poles:
   Support columns, three 2" x 6" laminated and precut at factory

12' truss and column spacing

Stall Framing:
   12' x 12' stall wall panels 8' high, tongue-and-groove boards precut at factory

   Designs include solid walls, vented with air space between boards above 5' or solid to 5' with ½" diameter pipe grill

   24' cross partition sealing stall area from storage area

Roofing:
   Kynar 500® roof steel with ½" Heavy duty Thermax-R-insulation

   2' 6" cupola with electric fan

   Three 3' x 3' skylights with vapor barrier

   1' overhang on all sides

   Aluminum gutters

Siding:
   Kynar 500® finish steel siding with exterior steel wainscot

Doors:
   Two 4' x 8' sliding stall doors, 2" x 8" tongue spaces

   One 10" X 8" Raynor® overhead door

   One 3068 9-lite aluminum walk-through door

   One 12' x 9' sliding wall/door

Windows:
   Two 4' x 3' single sliding windows w/screens and shutters

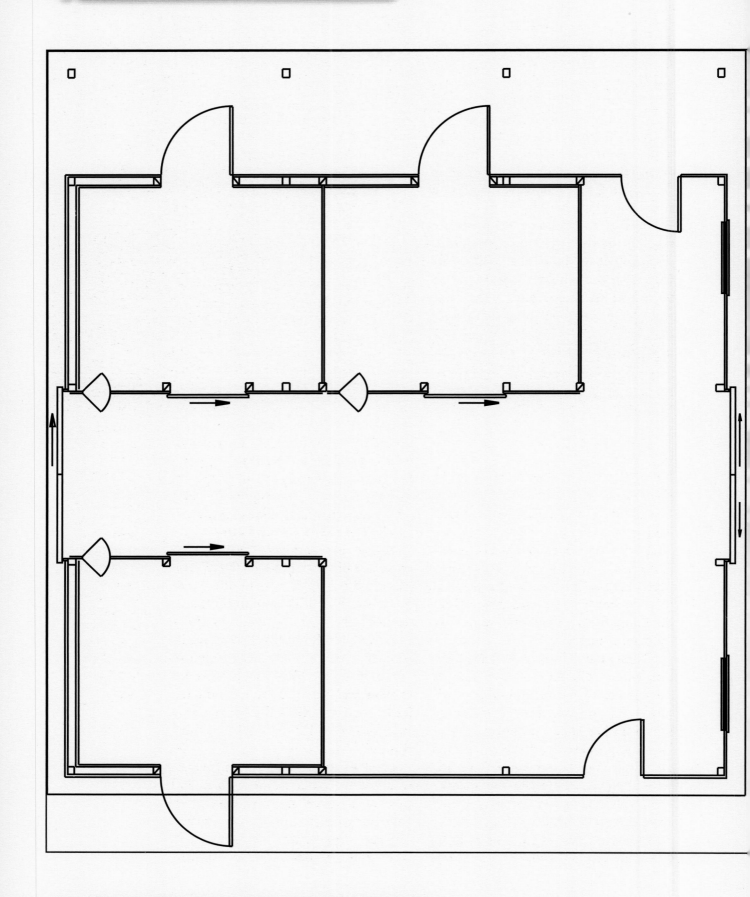

TWO-STALL ADDITION TO HOUSE

Description

Described by original owner Christa Panayo of Glendie Farm near Fredericksburg, Virginia, as suitable for senior citizens who want to take care of their horses themselves for as long as possible, this two-stall barn is attached directly to the house for minimal labor and maximum enjoyment.

Christa wanted to cut her barn labor time and have her horses close to her as she enjoyed their companionship. This functional and practical plan allowed her to do her labor without hiking all over the property, and she could step out to give her horses a quick pat any time of the day or night. The current owners, Bill, Ling and Kaja Baum, also enjoy this convenience.

The barn was designed and built by Col. Laurence Gardner of Falmouth, Virginia, and is basically a pole barn with 4" x 6" posts as functional supports. Box beams are tied to the support poles, making the roof self supporting and giving an open, airy appearance from inside. The rafters are on 2' centers.

The house and two-thirds of the barn are built on a 4" concrete slab. You can build this barn next to any house plan you like with minimum modifications. Be sure, however, to check with your local building codes and insurance carrier about firewall requirements. It is quite likely that a two-hour fire wall will be called for, meaning the barrier can hold back the flames for at least two hours before the flames burn through. This can be accomplished with either a wall of 8" cement blocks, concrete or fire-resistant lumber and two layers of 5/8" fire code sheet rock, depending on your local requirements. (Be sure you verify your plans with your building code inspectors and in-

The Baums have combined home and horse life smoothly with this two-stall attachment to her country home. Placed on the back of the house, it's a low-key way to add to the versatility of the acreage and keep her equine friends close at hand. (Glendie Farm, Falmouth, Virginia) Photo by M. F. Harcourt.

Dimensions:
36' x 24'

Poles:
4" x 6" set on concrete

Framing:
Ridge plate 2" x 8"

Rafters 2" x 6" on 2' centers with 2" x 4" collars

2" x 10" support girts run along top of tack room and stall front walls, bolted to outer front and rear walls and supported above by 2" x 6"s anchored to rafters

Stall posts are 4" x 6" with 2" x 6" support along the top and another flat on the top of that. "Y" supports from these are 2" x 6" up to rafters

Sill girt/skirt boards are 3 courses 2" x 8" tongue-and-groove pressure-treated lumber

Roof:
¾" chip board or plywood

15 lb. felt

215 lb. shingles

Siding:
T1-11 exterior ply

Exterior doors:
Pair 6' x 8' sliders for barn exit with galvanized rollers, T1-11 with particle board attached with 1" x 2" framing on ends

1" x 6" decorative boards on outside, and inside surfaces have 12" galvanized metal splash board since aisle way is also wash area. Alternately, use treated lumber for splash area

Trim:
Gutters on eaves with downspouts

Eaves 1" x 8"

1" x 6" fascia

1" x 2" finish boards

Windows:
3' x 21½" set in center of 12' increments

Outside trim for each, 1" x 4" finished boards

Interior trim for each, 1" x 6" rough-cut oak to match stalls

Stalls:
Inside stalls 1" x 6" rough-cut green oak with 2" x 6" cap on front of divider walls. Stall divider is central 4" x 4" with 1" x 6" lumber stacked on either side, 2" x 6" cap. Walls approx. 4' 4" high with no upper or front grills or restraints

Tack Room:
Studs 2" x 4". T1-11 on aisle side and in tack area, storage area unfinished. Tack room faced top and bottom with finished 1" x 4"s

surance carrier before you begin any construction.)

The third of the barn where the stalls were located was left with a dirt floor. A layer of gravel was put down, followed by 2" x 4" boards set on their edges, then a layer of 2" x 8" oak boards over the top to give the stalls oak floors. The 2" x 4" boards are treated, but the oak boards are not, as oak will cure better for this use if left untreated. The 2" x 4"s are set at 18-inch intervals with the 2" x 8"s laid across them. At each intersection, three spiral nails were used to se-

cure the joint. Builders used spiral nails so that as the horses walk on the floors, the nails don't work their way upwards. Ring shank nails would probably be even better for this purpose as they get a good "bite" into the wood each time they sink. Experienced builders tell us that screws are your best bet to anchor the boards. Whatever you use, inspect fasteners periodically to ensure a smooth, safe surface.

While early barns in the United States often had oak floors, especially in areas with German settlers who used this wood in European barns, many horsemen today have never seen a wooden floor in a stall. The older the barn and the further north you go, the more likely you are to find wooden stall floors.

Belmont Race Track in Belmont, New York, has barns with wooden floors, and this type is quite easy to clean and similar to stall mats. They are not without, drawbacks, of course: They can be slippery if not bedded adequately, and you must check for rot or insect damage through the years to prevent weak spots from developing. To prevent rot, the Baums pick out the stalls two or three times a day and clear up any damp bedding, sweeping the damp areas completely clean before laying dry shavings back down.

The Baum's tack room and office combination gave them a place to plan their activities in comfort as well as to store tack without mold problems—the area is serviced by the house's heating and air conditioning units. With adjoining structures, adding the necessary ductwork was reasonably inexpensive during construction.

Next to the office/tack room is a larger area that serves as a feed room and general maintenance room. Here the Baums not only keep feed stored in an old chest freezer for freshness and rodent control, but barn implements, hand tools and barn and garden supplies all stay here as well.

Since the Baum's horses have been together for years and get along well, the Baums use only a half-wall partition between them. This allowes them plenty of socialization while stabled, and went a long way toward giving this small barn a larger, more open and airy look.

No skylights can be found here, as they would add heat to the small area during the summer, but there is a large front door, two good-sized side windows, and a window over the front door

The closed doorway (through a fire-resistant door) leads directly into the house, while the door with a side glass panel opens into a combination tack room/office that contains the hot-water heater for the house and barn. Due to its close proximity to the house, this area is serviced by the heating-air conditioning system for the house, helping preserve tack and other leather goods. (Glendie Farm, Falmouth, Virginia) Photo by M. F. Harcourt.

for light and ventilation. An added metal vent over the aisleway helps to take out the rising hot air, and individual stall fans keep things moving on still, hot days. The spacious center aisle of roughened concrete was sloped to the outside and doubles as the wash area. For such barns, a fall of 1" to 4" will suffice to adequately drain the area.

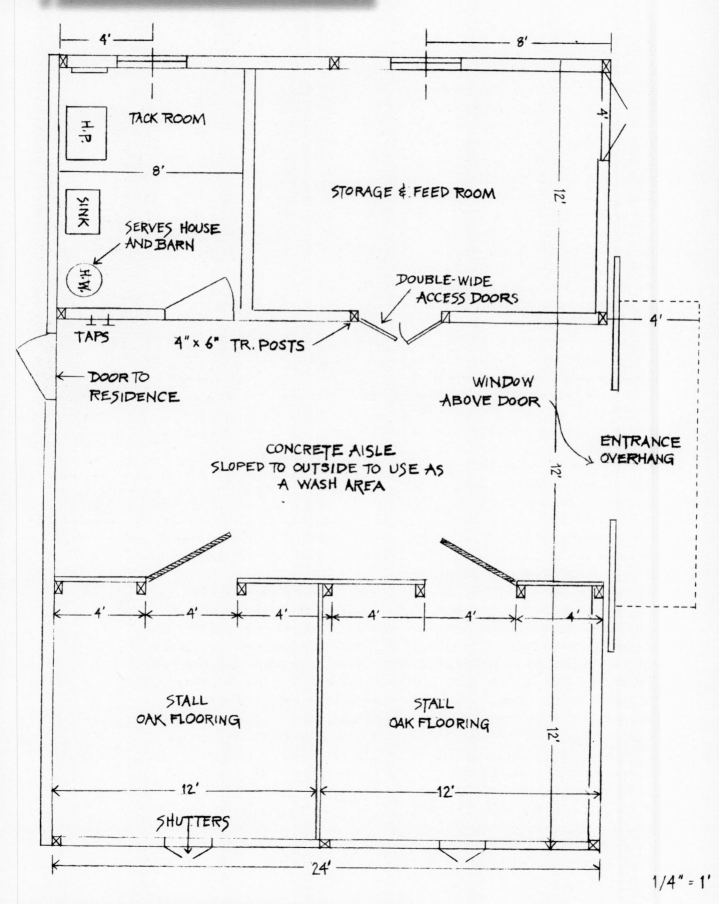

4'

8'

4'

TACK ROOM

H.P.

8'

SINK

12'

STORAGE & FEED ROOM

SERVES HOUSE
AND BARN

H.W.

DOUBLE-WIDE
ACCESS DOORS

TAPS

4" x 6" TR. POSTS

DOOR TO
RESIDENCE

WINDOW
ABOVE DOOR

ENTRANCE
OVERHANG

CONCRETE AISLE
SLOPED TO OUTSIDE TO USE AS
A WASH AREA

4'

12'

4'   4'   4'   4'   4'   4'

STALL
OAK FLOORING

STALL
OAK FLOORING

12'

12'   12'

SHUTTERS

24'

1/4" = 1'

FOUR-STALL WITH RUN-IN SHEDS

Description

Between the first and second editions of this book, Capt. and Mrs. Douglas Williams moved from Chesapeake, Virginia, to the rolling farmland of Surry County, Virginia. Designed by the Williams' and one of the authors of this book, their new barn on Chestnut Hill farm was built by Bryant Sherrill of Chesapeake, Virginia, who also made invaluable suggestions and modifications. The Williams incorporated a number of them into a low-maintenance barn that saves on footsteps, ladders, storage and resources.

Combining the formal stable area with run-in sheds under the same roof has been effective in saving building materials (shared back and side walls) and daily maintenance. Rather than hiking far afield to drop hay in sheds spread over the 20 acres of Chestnut Hill, the barn provides a one-stop arrangement. Doug Williams can drop hay into the stalls from the overhead loft, or kick a bale out the end door and just carry it around the corner to the run-in sheds, each of which services a different field. Like an octopus' arms, the surrounding pastures reach alongside one another to meet at the barn/shed combination.

The Williams' horses are used for foxhunting throughout the winter, and are turned out during the day and brought in at night. In the summer, the horses are off work, turned out at night and kept in during the day as heat and flies dictate. The fencing throughout the facility is Gallaher New Zealand stainless steel wire with a ½" polyethylene woven strip along the top.

To avoid the "1,000-pound termite" effect of horses roaming loose near the building, each wooden corner is edged with metal stripping. A pre-formed, heavy gauge strip fastened with screws is a good bet. No gutters or drain pipes

Combining comfortable stalls with large and small run-in sheds under one roof, this moderately sized barn saves on materials, resources and horsekeeping time. (Chestnut Hill, Spring Grove, Virginia) Photo by D. A. Williams

Dimensions:
36' x 48'

Poles:
6" x 6" pressure-treated posts

Framing:
Sills and plates 2" x 10" pairs

Rafters 2" x 6" on 16" o.c.

Joists 2" x 10"

Stalls:
2" x 8" tongue-and-groove partitions to 4', metal pipe grid 3" o.c. above

Roofing:
Fiberglass shingle, vented along western face with five metal static ventilators

Siding:
T1-11 siding with plastic "Z" strip at horizontal junctures

Trim:
16" fascia extension with 2" x 4" support under front overhang

Doors:
Front stall doors, 4' x 7' swinging Dutch doors, top sections 4' x 3' for broad, solid appearance and scale complementing house style

Heavy-duty strap hinges, sliding bolt latch at top and kick latch at bottom

Inside stall doors, 4' x 8' sliders with pipe grid section, hardware purchased as package through DT Industries, Inc.

End door, north, 6' x 8' sliding with 2" x 4" block frame for horse-stopper bar to drop into end door, south 32" x 80" standard solid exterior with 2" x 4" horse stopper swinging from bolt into right side frame, setting into site-built 2" x 4" bracket on left side frame. If this is to be a high traffic door, use a 36" width for equipment access

Floors:
Stalls have 45 2" x 4" boards set on edge over sand with crushed rock filler

Work areas have 4" concrete floor with roughened surface

Loft has two layers, offset, of ½" plywood

Windows:
Three standard 27" x 45" double-hung windows, two facing into run-in stalls, one from tack room to south side. ¾" electrical conduit pipes used for bars on exteriors, set in 2" x 4" framing

Lighting:
8' double fluorescent fixtures throughout stalls work areas and two in loft. Switches placed at each end of barn.

Nine double electrical outlets

Five halogen floodlights on exterior corners

Water:
Frost-free hydrant feeds hose reel in work area to service stalls and nearby field waterers Quick-recovery hot water heater,

lead into the pastures, as the Williams' believed that a concentrated stream into the sandy soil would lead to erosion. Instead, the roof drains off into a pea-gravel strip that surrounds the barn. To prevent rainfall from dripping into the seams of the plywood siding, a plastic drip-edge was installed where each sheet meets another horizontally.

An 18" cornice was also built onto the north and south ends of the roof.

The stable area is surrounded with a white three-rail fence separating the barn from the house and grounds. To keep dogs safely enclosed, a strip of electrified ½" webbing runs between the two bottom boards.

For afternoon shade in the largest

run-in, which faces west, a 6' x 12' sheet of agricultural shade cloth was hung across the opening. It is weighted at the bottom with a section of 2" x 4", attached only to the cloth and not to the door frame, so it can easily tear free if a horse is cast.

Grooming is handled in the stalls by choice, and no formal wash area was planned, as bathing the horses is not a frequent requirement for foxhunters. Instead an electric outlet is near each stall for quick vacuuming of each horse, and a small electric water heater in the feed room provides quick sponge-baths for touchup work. Nine outdoor-type double outlets are placed throughout the barn's stall and work areas for fans, clippers, vacuums, etc. The lighting is centered over each partition. The end stalls each have a fluorescent fixture against the outside wall, so grooming, clipping and tacking up can take place with no shadows.

Stall lights and work area lights are on separate switches. Obviously, lighting and electricity are priority items here, mostly because foxhunting is a winter sport, with plenty of early morning and late afternoon preparation and wrapping-up time, with little sunlight available.

Instead of leaving the aisle behind the stalls as an open storage/walkway areas, a pair of smaller rooms provide narrow but perfectly functional feed and tack storage areas. Since the management tools and saddlery require wall storage, the combination of smaller rooms and an access hallway doubles the number of walls available for hanging items. A built-in counter and cabinets in the small tack room take further advantage of wall space.

After spending years climbing vertical ladders, the Williams' found a gentle staircase a more attractive and safer idea. Now rather than storing only hay, which gets there via a movable electric conveyor, the loft can serve as general storage or even be converted to an apartment some day. The space taken up by the stairs, a 4' x 13' section of the work area, also provides under-stair storage or a dog house. The wall of that area offers yet another place to hang a

This north end view shows off the traditional design of the cupola along with the repeated "X" motif carried from the stall doors to the large end slider and the cupola sides themselves. Also visible is a small enclosure beside the large sliding door that houses a fire extinguisher, one of several in the barn. The gateleads to the pasture areas, which meet the sheds along the barn's west side. (Chestnut Hill, Spring Grove, Virginia) Photo by N. W. Ambrosiano

The south end of the barn houses a run-in shed, used now for general storage, and shows the barred window to the tack room and the access door to the central work area. Bars across both the walk-in door and the loft door above prevent people or horses from wandering through when the doors are open. (Chestnut Hill, Spring Grove, Virginia) Photo by N. W. Ambrosiano.

Standing in the large run-in shed with its shade-cloth screen, this mare enjoys some time out of her stall. Note the thin white strand extending from the door frame at right, which is the divider wire between two pastures. Such webbed electric wire options allow visible barriers that are easy to move back and forth (Chestnut Hill, Spring Grove, Virginia) Photo by N. W. Ambrosiano.

pitchfork, halters or whatever. The loft itself is made more useful by a knee wall, which raises the roof another 4' and reduces bumped heads dramatically.

The 12' x 12' stalls have the 2" x 4" wood/sand floor grid described elsewhere in this book, with 45 2" x 4"s set on 16" to 18" of rough construction sand and crushed rock between the boards. Bedding is placed directly on top, with no rubber matting needed. To further ensure a fresh barn, a cup of an aluminum sulfate product called Sweet PDZ is sprinkled in each stall periodically to eliminate ammonia odors.

Topping off the whole picture is a large cupola, 3' on a side with a steeply pitched and curving roof evocative of the grandstands at the Kentucky Derby site on Churchill Downs. As this barn was built to stand within 100' of a restored 1700s home, maintaining a consistent, complementary style was important. The barn and house are the same colors; pigmented oil stain was used instead of paint.

The staircase provides safe access to the loft for varied needs, and by enclosing the area beneath it a useful storage area is created. At this barn the resident Mastiff pair uses that area for their dog house, wandering upstairs to use the loft doorway as their observation post. (Chestnut Hill, Spring Grove, Virginia) Photo by N. W. Ambrosiano.

Liberal use of grillwork throughout the stall area offers good ventilation and light. Note the smooth door latch arrangement, made safe by pre-purchased door latches that operate a catch in the roller assembly above. No projecting pieces are near the door edge. On each stall door a homemade blanket rack is placed, made from electrical conduit and wooden end pieces. The projecting saddle rack to right of center is a trimmed, stained 4" x 4" with a hook attached, looped through a screw eye in the door. It can be turned and hung almost flush with the door when not needed. (Chestnut Hill, Spring Grove, Virginia) Photo by N. W. Ambrosiano.

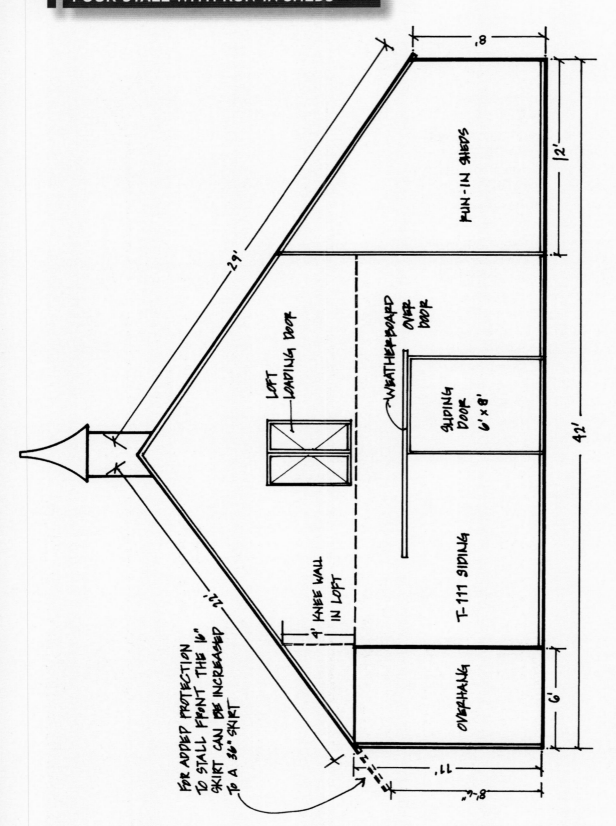

.8'

12'

RUN-IN SHEDS

42'

LOFT LOADING DOOR

WEATHERBOARD OVER DOOR

SLIDING DOOR 6' x 8'

29'

4' KNEE WALL IN LOFT

T-111 SIDING

FOR ADDED PROTECTION TO STALL FRONT THE 16" SKIRT CAN BE INCREASED TO A 36" SKIRT

OVERHANG

6'

11'

8'-9"

4 STALL BARN WITH RUN-IN SHEDS BUILT-IN

ELEVATION

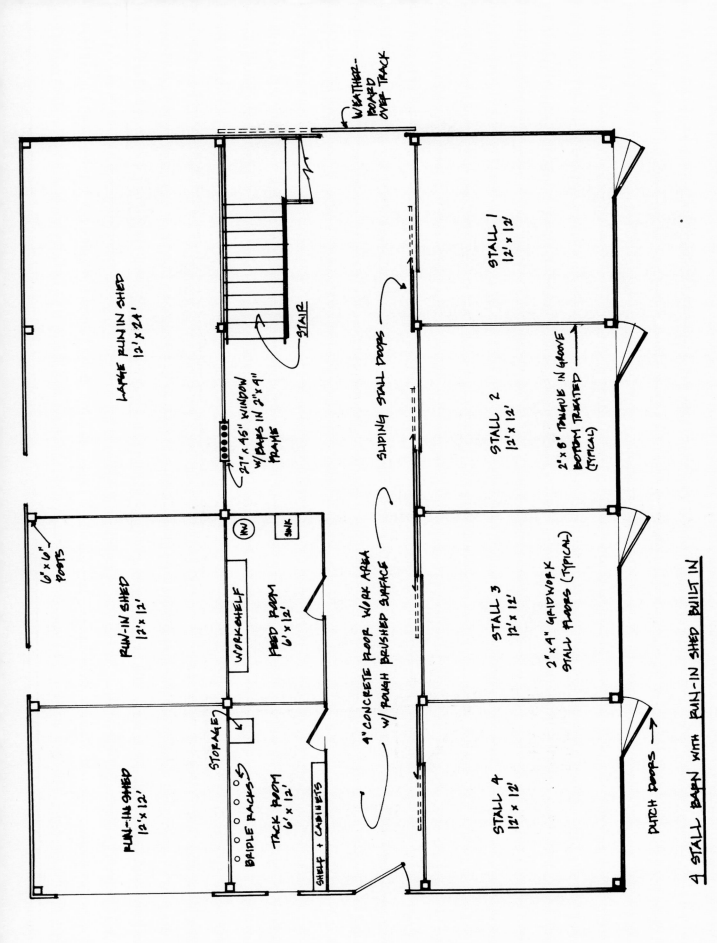

WEATHER-BOARD OVER TRACK

LARGE RUN-IN SHED
12' x 24'

STAIR

27" x 45" WINDOW
w/BARS IN 2" x 4" FRAME

RUN-IN SHED
12' x 12'

6' x 6" POSTS

WORKSHELF

FEED ROOM
6' x 12'

HW

SINK

STORAGE

RUN-IN SHED
12' x 12'

BRIDLE RACKS

TACK ROOM
6' x 12'

SHELF + CABINETS

SLIDING STALL DOORS

STALL 1
12' x 12'

STALL 2
12' x 12'

2" x 8" TONGUE IN GROOVE
BOTTOM TREATED (TYPICAL)

4" CONCRETE FLOOR WORK AREA
w/ ROUGH BRUSHED SURFACE

STALL 3
12' x 12'

STALL 4
12' x 12'

2" x 4" GRIDWORK
STALL FLOORS (TYPICAL)

DUTCH DOORS

4 STALL BARN WITH RUN-IN SHED BUILT IN

RED BRICK BARN

Description

This stunning red brick barn was built exclusively for Mr. and Mrs. Douglas Dyke of Rockville, Virginia, by O. Earl Hayes of Richmond, Virginia. Mr. Hayes, a longtime friend of the Dykes, wanted to build a barn that would complement the home he had constructed for them.

With 24 acres of land at their disposal, the Dykes decided to build the home on one 12-acre section and place the barn and pastures on the other 12. This cinderblock with red-brick exterior barn has coin corners and jack arches to match the home and is so well built and attractive that it could easily be turned into a handsome home in the future. While the initial investment in such a facility is large, the number of potential future uses help compensate for the initial expense.

Mr. Hayes chose a design from the farm where he rode horses as a child, then added amenities reflecting modern advances. Ample ventilation flows through the wide windows and doors, and the overhead loft acts as a massive ridge vent. The combination of cinderblock and brick walls gives a cool interior in summer and helps hold interior heat from the horses in the winter. Masonry walls should be sealed to reduce moisture. Footing throughout the building is bluestone chips, and the walls were finished in rough-cut oak to shield the horses from the hard cinderblocks and to offer an attractive interior.

Many horsemen find that a wooden wall over brick or cinderblock is helpful in protecting a kicking horse against capped hocks, skinned joints and even possible fractures. While nothing short of full padding will protect every horse, steps such as wooden shielding can make solid walls more forgiving.

Access to the loft area is by a staircase, rather than a simple vertical ladder. For general safety, as well as convenience, this is an excellent item to plan for and makes your loft far more versatile for general storage.

MATERIALS

Dimensions:
65' x 58' 1"

Outer walls are 6" cinderblocks with 4" red brick covering. Should have 1" air between, adequate Durowall and vertical reinforcing bars

Interior Framing:
Poles: A combination of 6" x 6" and 4" x 6" interior supports set 12' on center, sitting on 24" x 24" concrete footers

Loft floor joists are 2" x 10" set 16" o.c. supported by a 12" x 12" wood beam on 6" x 6" posts with concrete footers

Garage door openings are framed with two 2" x 4" jacks to support the paired 2" x 10" boards that form the header

Loft Floor:
¾" plywood sheeting

Roofing:
½" plywood with roofing felt and 240 lb. shingles to match the home

Exterior:
4" red brick over 6" cinderblock

Trim:
Coin corners, jack arches over windows and doors. Gutters on eaves with downspouts all around building, including rear storage shed

Exterior doors:
10 garage doors at each end. For best fire safety do not use overhead, rolling garage doors. Use standard sliding or swinging doors.

Windows:
Stall level, 3' x 3' with shutters

Loft level, 2' x 3' 2"

Interior:
Stalls: finished in rough-cut oak 1' x 6"

Tack room: paneled with T1-11 plywood over insulation, floors ¾" plywood with vinyl tile

Office: birch paneling over sheetrock and insulation, floors ¾" plywood with vinyl tile

The rear view shows the roomy equipment storage area, 16 feet deep, that can accommodate tractors, mowers, jumps and more. The far right bay has been framed in to provide lockable, secure storage for valuable items or things that need more weatherproofing. (Bill & Lou Webb, Rockville, Virginia) Photo by M. F. Harcourt.

Solid, simple and elegant, this brick barn combines ample space for horses, storage and equipment with a raised loft area and attached tractor shed off the back. (Bill & Lori Webb, Rockville, Virginia) Photo by M. F. Harcourt.

Early in construction: Here you see the cinderblocks going up doorways and windows ready for sills and supporting members for the bricks to come above. (Bill & Lou Webb, Rockville, Virginias) Photo courtesy of Miriam Dyke.

Walls are up, with one end bricked in. The rafters are in place and the central hayloft area has been set up. (Bill & Lou Webb, Rockville, Virginia)

Before the plywood's laid for the roofing, the framing of the equipment lean-to must start. Notice that the pitch of its roof varies from that of the main building, in order for the front edge to be high enough for a working clearance. (Bill & Lou Webb, Rockville, Virginia)

Last after the trim work and inside stalls come the roofing for both the main barn and the loft area. (Bill & Lou Webb, Rockville, VA) Photos courtesy of Miriam Dyke.

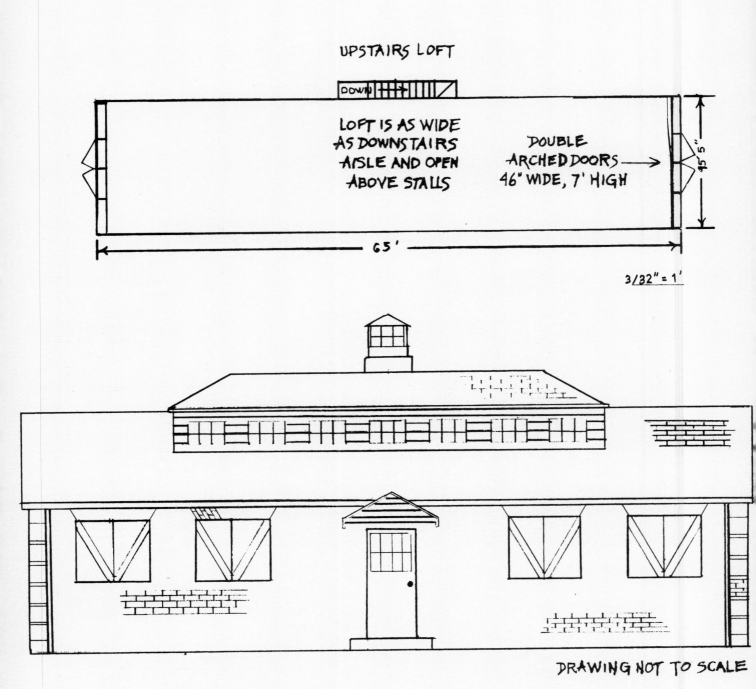

UPSTAIRS LOFT

DOWN

LOFT IS AS WIDE
AS DOWNSTAIRS
AISLE AND OPEN
ABOVE STALLS

DOUBLE
ARCHED DOORS →
46" WIDE, 7' HIGH

45' 5"

65'

3/32" = 1'

DRAWING NOT TO SCALE

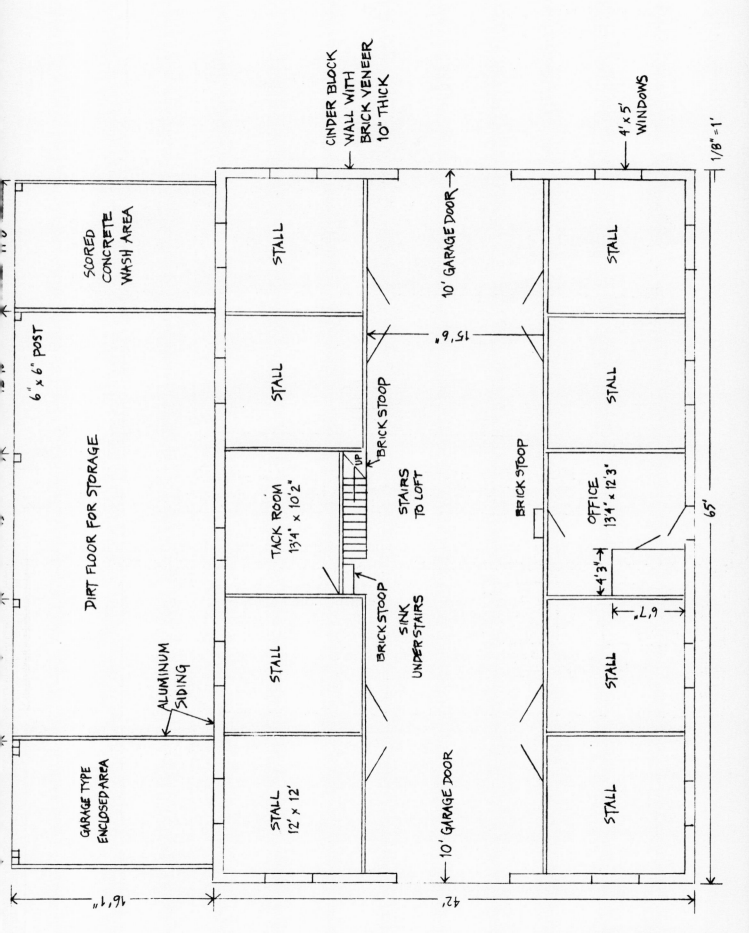

CINDER BLOCK WALL WITH BRICK VENEER 10" THICK

4' x 5' WINDOWS

1/8" = 1'

SCORED CONCRETE WASH AREA

STALL

STALL

10' GARAGE DOOR

STALL

STALL

15'6"

6" x 6" POST

DIRT FLOOR FOR STORAGE

TACK ROOM 13'4" x 10'2"

BRICK STOOP

UP

STAIRS TO LOFT

BRICK STOOP

OFFICE 13'4" x 12'3"

4'3"

65'

ALUMINUM SIDING

STALL

BRICK STOOP

SINK UNDERSTAIRS

9'7"

STALL

GARAGE TYPE ENCLOSED AREA

STALL 12' x 12'

10' GARAGE DOOR

STALL

16'1"

42'

This magnificent barn houses 18 horses and allows a circle of jumps around its covered exercise track. An inviting central service area features tack and feed rooms and a wash stall, and a central aisle with doors at either end adds to the ventilation and ease of use. This structure was developed and erected by Upperville Barns of Upperville, Virginia. (Miye Cayton, Warrenton, Virginia) Photo by M. F. Harcourt.

ROUND BARN

Description

Round barns, no matter where they're found, are always attention getters. The handsome model on page xx was designed with indoor jumping in mind. Former owner Mrs. Jeffrey Babcox rode a circle of jumps in her field to get the size she preferred, then had her father, Mr. John Breeden, sketch out the basic idea. Working with Upperville Barns of Upperville, Virginia, Breeden planned an inner circle of stalls with an outer exercise ring of bluestone chips that is over 90' in width and has plenty of headroom for a rider. An aisleway pierces the stall circle, allowing service areas (tack, feed, etc.) to be centrally placed and a storage/viewing area to be located above the service circle.

Besides providing exercise room, the design allows a bit of privacy in a small boarding operation, as horses are distributed outward around the circle diminishing the visual impression of an aisle full of animals being groomed or handled. The central tack/feed/washing area gives a warm, friendly core to the operation, giving riders the option of community time in the center or more privacy out by the stall doors.

While the model shown is built with wooden support columns and rafters, the roof system could be constructed in a number of ways. Steel girders or a combination of metal and wood could be used, depending on what's available in or recommended for your area. As with most round barn designs, this is not a perfectly round building: It is built as a circle of 10-foot panels, thus allowing use of regular flat lumber and stock windows. The stalls vary widely in their sizes, with none being square and some having a rear wall as short as 6' paired with a 10- or 12' front wall.

MATERIALS

Dimensions:
    Diameter 95' 6", circumference 301'

Exterior:
    Walls: 3 courses of 8" cinderblocks set on 12" x 16" concrete footers with wooden work above

    Outer framing: 4" x 6" posts with 1" x 8" finished boards lapped for a 6" exposure

Trim:
    1" x 8" soffits
    1" x 6" fascia

    No gutters—gravel layers alongside building allow gradual runoff. Use perforated pipe with 1" to 8' fall under gravel for more drainage

    Building sits on crown of hill for ease of construction and good drainage exterior

    Doors: Two double sliders 10' each to match exterior

Roof:
    1/2" CDX plywood

    15 lb. roofing felt

    fiberglass shingles

Interior:
    Stalls: 18 stalls varying from 14' x 14' to 10' x 12' with bluestone floors

    4" x 4" brace posts set on 14" x 18" concrete footers placed at varying intervals due to round shape and unique interior layout.

    Walls: 2" x 6" tongue-and-groove, 53" high with iron grill on all sides.

    Stall doors 4' wide sliders, 2" x 6" tongue-and-groove, grill above 53" level.

    Wash pit: 9' x 12' with roughened (barn broom finish) concrete floor, same walls as stalls

    Feed Room: 8' x 12' with concrete floor, same walls as stalls

    Tack Room: 17' x 12' with sink and commode sectioned off, concrete floor finished in vinyl tile, walls done as stalls, with paneling or sheetrock

    Track/Aisleway: 12' wide, bluestone chipped footing, with free overhead clearance from exterior wall to first interior stall support pole. To withstand high traffic on the track, underlay 1½" of bluestone with 4" of gravel

A bewildering array of rafters, beams and plates criss-crosses the upper reaches of the barn. The narrow grid of railings at top center surround the flooring of the "conning tower" or second layer on the cake. Ventilation fans are visible hanging from one beam. Photo by M. F. Harcourt.

Even with the curvature of the walls in this round barn, doors can be constructed to present no problem in allowing access to equipment, including a tractor-and-manure-spreader combination. Each of the two sections of the sliding doors is hinged, allowing them to follow the curved track of the walls. (Miye Cayton, Warrenton, Virginia) Photo by M.F. Harcourt.

The spacious aisle is over 90' in diameter, lit all the way around by sliding windows, shown, and plenty of large double-bulb fluorescent fixtures set between the rafters. As apparent here, the barn could be quite busy with horses being tacked up at each stall, and yet there would be no visible crowding. Standard sliding doors and fixtures are used throughout. Photo by M. F. Harcourt.

It is imperative you seek professional help should you decide to build a structure such as this one, with a unique roof design. You'll need assistance in designing and building a structural roof support system that will not only support itself but also whatever your climate may throw at you.

Because we feel that the roof support system can be done in several ways—all of which demand professional design and installation to be done correctly—

we have not included a diagram of the barn's support system in this book.

Talk with your architect or builder as to the best way to build this barn for your area. Need we reiterate the necessity of working with local inspectors and codes in every phase? While expensive to build, this barn may have a place in the industry or at least in many of our dreams.

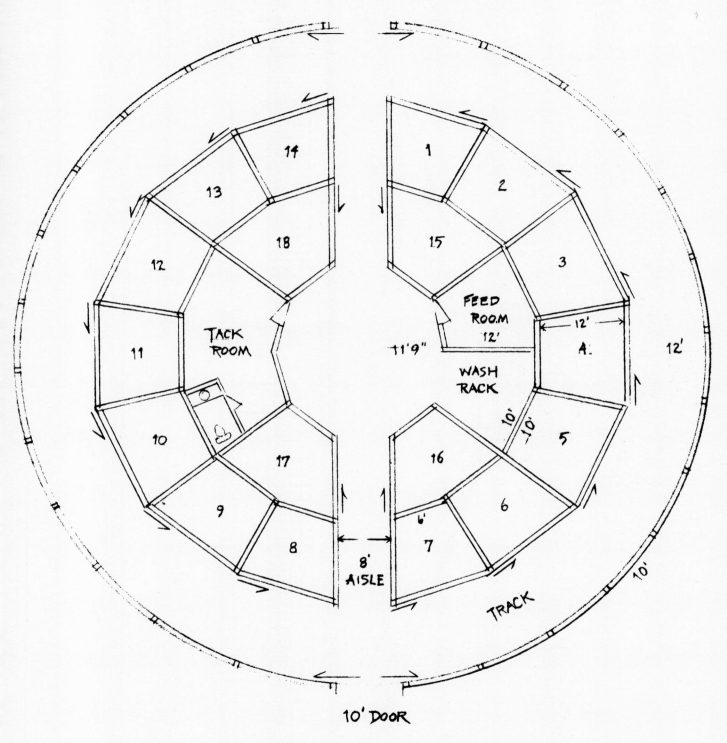

TACK ROOM

FEED ROOM 12'

WASH RACK

11'9"

12'

12'

10'

10'

8' AISLE

6'

TRACK

10'

10' DOOR

CIRCUMFERENCE = 301'

DIAMETER: 95'6"

1/16" = 1'

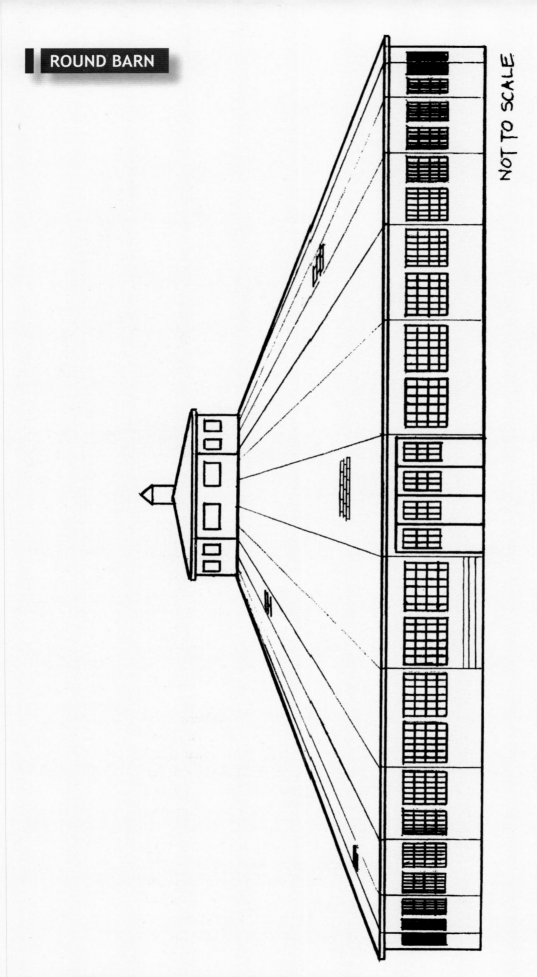

NOT TO SCALE

OVER-AND-UNDER HUNT BOX

Description

When deciding to build a second home in a warmer climate than their New England farm, Carl and Geraldine Lombard chose to build this house-over-barn in Southern Pines, North Carolina. The combination of good winter riding and a horse-oriented community has made the area increasingly popular. The Lombards selected local barn designer Linda Dreher to provide a plan that was both functional and attractive. Houses over barns are becoming more popular as a way to make horse keeping easier and reduce cost by letting one roof serve for both the home and barn structure.

Some people are choosing this type of structure as their preliminary living space as they plan to build their farm in stages. Having the barn complete with a preliminary living space until a larger home is built allows them to be close to their horses and supervise the construction of the detached home. When the larger detached home is complete this structure then becomes guest quarters, rentable space or living quarters for employees. Others choose a combination house/barn as a second residence, investment property or future retirement home.

In designing this house/barn combination, Dreher fused form and function to create a basic design that could be adapted into numerous interior layouts or exterior facades to meet the individual needs and taste of a variety of owners.

The Lombards have chosen cinder-block below with wood above as the basic construction materials for the structure. A lower level foyer serves as a downstairs entrance for the home as well as a barn entrance and tack room.

With the upper deck and rear stalls facing the paddocks, woods and a valley, the Lombards can enjoy privacy, and a lovely view, knowing that care of their horses takes minimum effort. (Carl and Geraldine Lombard, Southern Pines, North Carolina) Photo by M. Harcourt

It's always hard to envision how the finished product will look. But once completed and landscaped, the entrance to the Lombards' home/barn will set the tone for the style that they will create for their farm. Photo by M. Harcourt

MATERIALS

This barn can be built in many different forms: cinderblock for both Stories; cinderblock under lumber; all lumber; or lumber frame with brick exterior, so no material list will be provided for this plan. Consult your architect or builder after you decide what type of structure you plan on building and have him or her provide you with a materials list that meets your specifications.

On one side of the foyer is a feed storage/work room. On the other side is a fourth stall. With a wide center aisle and three stalls on the back of the structure, there's room for four horses, or three horses plus hay or bedding storage.

Since there is a wonderful view overlooking the woods and a valley, the home has a deck to the rear that opens to a great room/kitchen/dining room flanked on each side by master bedrooms, each with its own master bath.

As a unique touch, the Lombards used a rough cut beam over the tops of the downstairs windows. That, coupled with a stucco exterior on the lower cinder blocks, gives this home/barn combination a French country feel. But each builder can put his or her own touch on the style and ambience of the building, depending on what the builder chooses for exterior materials, fencing and landscaping.

While this home/barn may appear to be small, good design of space affords ample comfort for both human and horse occupants.

The owners of StarWood Farms utilized the same basic design as the Lombards for their home/barn but added a fourth stall on the back side, a carport to the front left and a covered porch to the right that serves as an upper deck/side entrance to the home above. With red brick over the cinderblocks/wood building materials a home-like appearance was created on the front of the structure, hiding the fact that it also serves as a home for horses. StarWood Farms, Southern Pines, North Carolina (Photo by M. Harcourt)

When the StarWood Farms barn is viewed from the side/back, it's evident that horses also call this lovely structure home. (Photo by M. Harcourt)

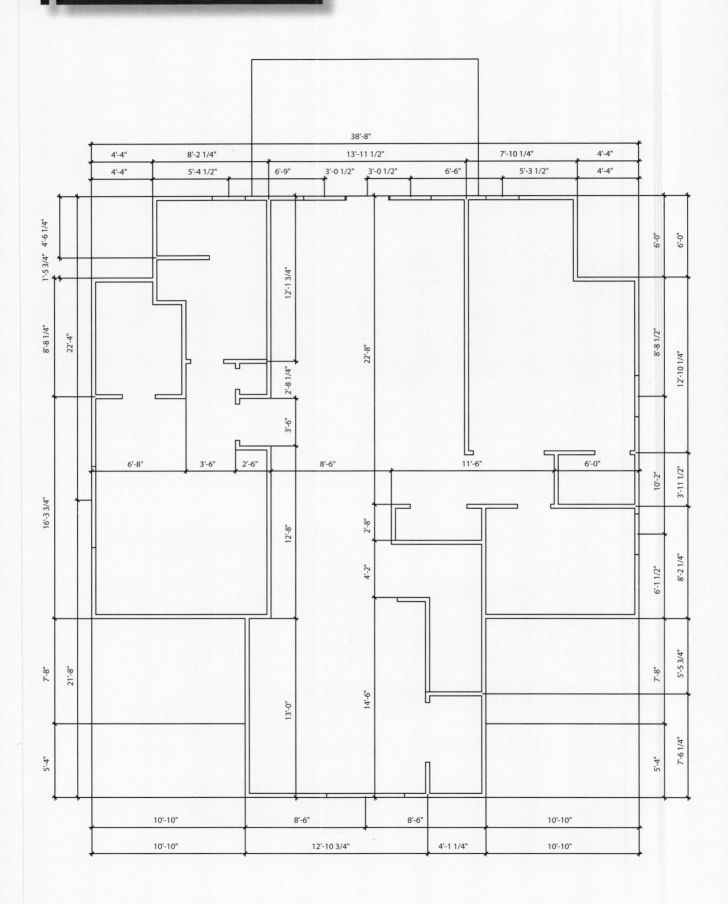

A side access garage door provides a wide opening for unloading supplies directly into the storage / work space. The smaller access door makes it convenient to do daily chores. Kirkland Barn, Ashley Heights, North Carolina (Photo by M. F. Harcourt)

TWO-STALL MULTI-PURPOSE BARN

Description

Lynn and Sammi Kirkland took a simple barn plan and designed it to meet their own unique needs. Sammi uses part of the barn to raise her fancy chicks before turning them out into the larger chicken yard attached to the back of the stalls. By using the barn as the focal point of their operation, the Kirklands can service all their animals easily from a common feed and storage area. By dividing up the pasture, horses and goats can be rotated and the chickens can have some fresh grass by utilizing some of the pasture area also. This simple yet effective plan can easily be adapted to meet whatever needs you might have. With a garage door for easy access of feed and equipment and some thoughtful planning, this little barn can provide a wealth of space and arrangement to meet many of a family's farmette needs.

MATERIALS

12	6"x6"x16"
8	4"x4"x16'
17	80 lbs. Quickrete
40	2"x6"x12 treated
57	T 111 siding
23	1"x6"x12' trim
13	¼" plywood
2	36" Exterior doors
1	7'x9' Garage door
2	windows
120	2"x6"x 12' yellow pine or spruce
6	2"x10"x12
40	Hurricane straps
30	2"x6"x8 R.C.
20	2"x6"x16'
60	2"x4"x12'
26	3'x14' metal roofing
40'	Ridge cap
1000	screws 2"
36	⁵/₄"x16' treated decking boards

Design and arrangement allows the Kirklands to provide shelter for horses, goats and chickens plus storage for all the necessary feed and supplies to take care of their animals. Good planning and a functional arrangement of the interior will allow you to utilize the space to fit your specific requirements. Kirkland Barn, Ashley Heights, North Carolina (Photo by M. F. Harcourt)

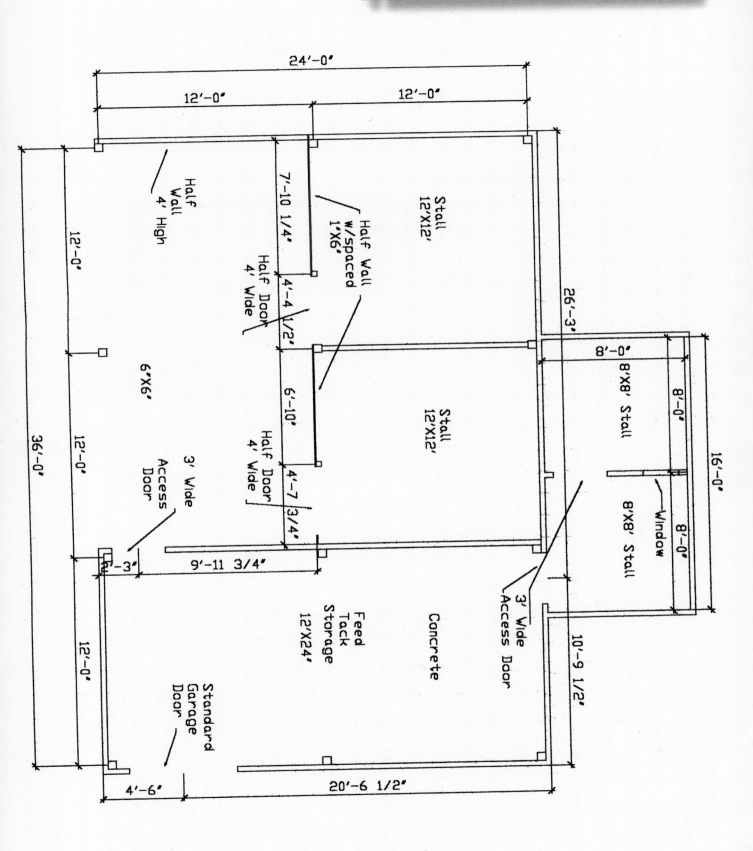

This attractive and compact barn provides stalls, tack room, hay storage, pasture shelter and free choice access to hay all under one roof. Ideal for the two horse family, the Montgomery's barn makes keeping horses simple and efficient. (Ken and Helen Montgomery, Ashley Heights, North Carolina) Photo by M. F. Harcourt

UPSCALE TWO-STALL BARN

Description

Small in size but large in function describes this 36' x 36' two-stall barn built by Old South Building Systems for Ken and Helen Montgomery of Aberdeen, North Carolina.

The Montgomery's combined their love of horses and Border Collies by building a barn that serves as a home for their two horses, as well as tack, hay and general supplies needed for their flock of Barbados sheep, which are used to train the Border Collies.

With two stalls that open to a 12-foot overhang for protection from weather and the hot North Carolina summer sun, plus access to a 12' center aisle, Helen can manage her horses in any of a number of ways. Add to that a run-in shed across the entire end of the barn used to hold a round bale holder allowing the horses constant access to hay all winter

MATERIALS	
84	2"x6"x12 treated
8	2"x12"x20'
146	2"x6"12 yellow pine
130	2x6 Truss hangers
18	2"x6"x12'
92	2"x4"x12'
63	2"x6"x12' Tongue & groove
18	2"x6"x8' Tongue & groove
45	T111
2	2"x8"x12
2	36" Exterior doors
16	1"x6"x12
4	1"x12"x12
16	1"x4"x12
26	22'x3' metal roofing
2000	metal screws
40'	Ridge cap

Hidden from the road view, the back of the Montgomery barn shows how a round bale can be fed free-choice to their horses, providing shelter and a constant source of roughage. With only two horses and a small donkey vying for space, a section can easily be blocked off to provide seasonal equipment storage. (Ken and Helen Montgomery, Ashley Heights, North Carolina) Photo by M. Harcourt

and you have a barn for all seasons and all situations. Interior hay storage for square bales for horses and sheep plus a 12' x 12' tack room complete the ample but utilitarian arrangement of this compact yet well-organized barn.

When planning your barn, take the basic size that meets your land and cost availability, then work on the interior arrangement to meet your specific requirements and needs of your horses, or any other hobby activities you, your family and friends may enjoy.

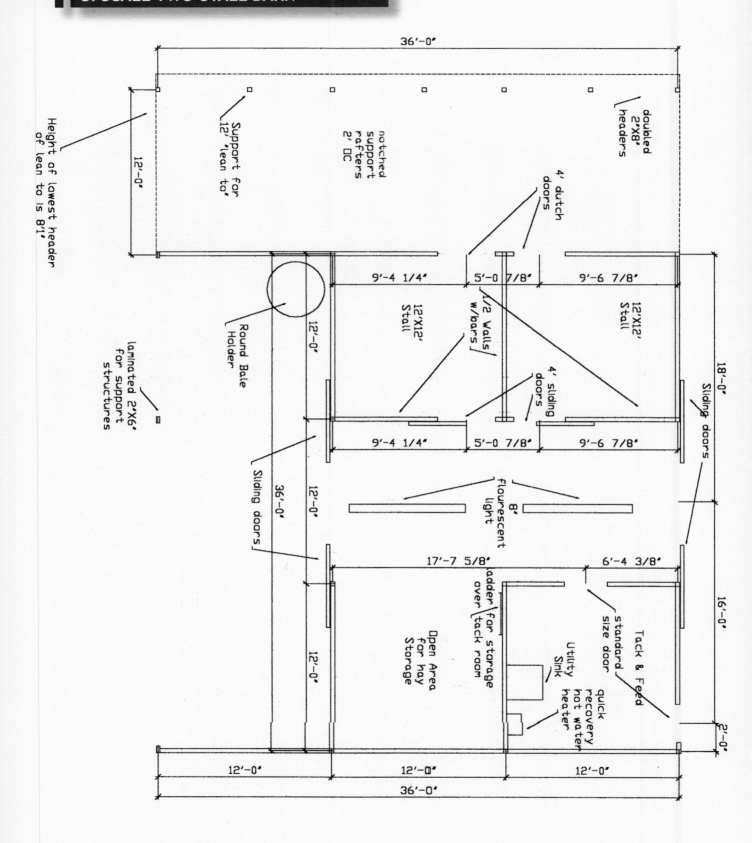

36'-0"

12'-0"

doubled
2"X8"
headers

Support for
12' lean to

notched
support
rafters
2' OC

Height of lowest header
of lean to is 8'1"

4' dutch
doors

9'-4 1/4"   5'-0 7/8"   9'-6 7/8"

12'-0"

12'X12'
Stall

1/2 Walls
w/bars

12'X12'
Stall

18'-0"

Sliding doors

Round Bale
Holder

laminated 2"X6"
for support
structures

4' sliding
doors

9'-4 1/4"   5'-0 7/8"   9'-6 7/8"

8'
flourescent
light

Sliding doors

36'-0"   12'-0"

17'-7 5/8"   6'-4 3/8"

12'-0"

ladder for storage
over tack room

standard
size door

Open Area
for hay
Storage

Utility
Sink

quick
recovery
hot water
heater

Tack & Feed

16'-0"

2'-0"

12'-0"   12'-0"   12'-0"

36'-0"

PRE-FAB INNOVATION

Description

Occasionally a new idea comes along that incorporates some existing but upgraded technology. Such an idea is being utilized by Cover-All Building Systems to produce reasonably priced yet durable alternatives to conventional barn roof building systems.

Cover-All Building Systems, Inc. is a manufacturer of steel-framed, fabric-covered structures that can be used for anything imaginable. The buildings are pre-engineered in clear-span widths from 18 to 160 feet. It didn't take long before horse owners saw the potential for a barn or indoor arena, and a new type of equine building system was born.

Available in a variety of roof line designs, the unique shapes allow snow to slide off easily. Engineered to meet snow and wind loads in each building

location and tested under severe winter conditions, these structures are not only strong but easier and less expensive than traditional roof structures to erect. While the translucent properties of the cover allow high levels of natural light in, good ventilation and reflective properties keep temperatures moderate in hot weather. A North Carolina dealer has even adapted a structure to the hurricane-prone Kitty Hawk area of the North Carolina Outer Banks with no problems.

The DuraWeave covers, according to the manufacturer, "are virtually maintenance free, have an unmatched strength-to-weight ratio and a tremendous rip, tear and puncture resistance." The strength of the fabric is impressive: a 10' x 10' piece of DuraWeave fabric will safely support 14 tons.

The interior can then be finished any way the owner wants: stalls, indoor arena, or a combination of both. The translucent properties allow for incred-

With larger Cover-All buildings, all facets of your operation can be under one roof. Stalls to riding ring to office and viewing area can all be part of the interior layout.

With a variety of roof lines and floor sizes, the Cover-All building system offers some equine operations a cost effective way to add workable space that is easy to maintain. Connections to auxiliary buildings are simple to do, allowing owners flexibility in their over-all farm plan.

The Cover-All building allows large amounts of natural light to reach all corners of the structure, reducing glare and shadows without increasing the interior temperature as glass or plastic would.

ible amounts of natural light, reducing shadows and glare but without increasing interior temperatures. In addition, ventilation is exceptional and there is much less sound reverberation making for a quieter, more natural environment for working horses.

The structure offers a clean line of design in a variety of colors that blend in with the environment and is easy to maintain. The building can be anchored on a foundation that is low or constitutes half a wall allowing the owner a number of facility design options.

With dealers located throughout the world, prices will vary due to shipping and labor. Find a dealer in your area to compare price per square foot of the Cover-All structures and other equine building systems available to you.

While it may not be a structure design for every operation, the pre-fab enclosure definitely has a place in the industry for anyone needing a large, covered, cost-effective structure for any reason.

Remodeling

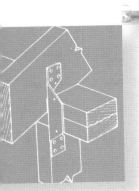

If you have bought land with older buildings already on it, you have the choice of remodeling or pulling things down. A discriminating eye that's open to opportunities is most useful here; even if the property was sold with buildings termed "horse facilities," there's no guarantee these structures are safe or sensible for your horses.

Even old cow barns with 50 years of thickly layered "history" can be useful, if you look things over as a possible starting point for a run-in shed, a tack room or even the core of a larger structure. Basically, you can fix up anything, given time and money. Your challenge is to come out ahead on both.

ASSESSING THE BARN

Some background reading on the subject of historic barns may help develop your eye for what to value and what to bulldoze. Three excellent books on the subject exist that will hone your perception of interesting old-fashioned barn construction and design so you can see worthy barns as the art form they are. *An Age of Barns*, by Eric Sloane (NY, Funk and Wagnalls, 1967) is one with lovely, detailed drawings of the tools, parts and completed buildings created by the early European settlers in North America. Two coffee-table books with large color plates are *Barn: The Art of a Working Building*, by Endersby, Greenwood & Larkin; and *The Barn: A Vanishing Landmark in North America*, by Eric Arthur and Dudley Witney (M. F. Feheley Arts Company Limited, Ontario, 1972).

Factors in a building's remodeling potential include the following:

✔ Few leaks in the roof

✔ Little or no water damage to the rafters or supports

✔ Straight roof lines

✔ Straight side walls

✔ Limited rot in main support structures

 Location suitable or at least reasonable for the new use

 Estimated cost of remodeling less than cost of new structure

There are some structures that because of their historic or sentimental value warrant remodeling at almost any cost. But this is a personal decision, with variables far beyond simple horsekeeping principles. Given a historic edifice with a distinct propensity for falling in a heap, perhaps you can develop alternatives:

■ Save the unique parts of the building, such as hand hewn rafters, pegged posts or weathered doors, siding and hardware, and incorporate them into a new building. This "new from old" technique is wonderfully effective and will keep a new barn in an older area from appearing out of place. Some old properties have scrap heaps with wonderful treasures hidden away that can give your new place infinitely more character. Line your tack room walls or grooming area with weathered siding, lay an antique door over a strong new frame or save old bricks for a handsome path to your tack or feed room doorway.

■ Prop up an old building shell with some solid framing and turn it into a fine run-in shed. Check carefully for old nails, wire or other hardware embedded in the wood or soil around the building before you turn your horses out through, as they're sure to find the most harmful ones in record time.

Checking your old buildings for rot or disintegrating underpinnings takes both a close look and a hands-on approach. With a crowbar, hammer and substantial screwdriver, you can tap, poke, pry and thump your way through every piece of structural lumber on the place. Looking over foundations, sills, support posts and beams, walls, and rafters, check for soft, disintegrating sections, dry rot or termite damage beneath the outer layers of wood.

Posts showing signs of rot can be replaced by jacking up the beam immediately beside the old post, cutting it out and inserting a new treated post in its place. If only a small part of the post is damaged, cut out the damage and add either a concrete base or cinderblocks, or lap-joint a new section of post in.

Weather or manure-damaged boards at the base of walls can be removed, the underlying soil dug out, and the boards replaced with concrete footers, cinderblocks or pressure-treated lumber above ground level. Then backfill the stalls and exterior with soil or gravel and pack it firmly into place. If extra measures to ensure drainage are needed, such as French drains or piping, get them in place before starting the backfill. Then finish with gutter, #6 gravel next to the building, splash zones and grading.

While you've got damaged boards off and soil dug out, add in electrical conduits or additional plumbing before you begin the finish work. And if as you remove sections you uncover some additional damage, don't succumb to the urge to just cover it over--it will resurface to haunt you when you paint or pave.

If you are adding to a building in a snow area, take into account the additional stress that occurs as snow masses in the juncture of the two roofs. If an older, higher barn has a lower "L" added, be especially aware of the strength requirements, as snow tends to build up at building joints and cause roof failures.

For interior layouts, you'll be fortunate if the support structures are built on 12' centers, which would allow useful 12' x 12' stalls throughout. As is likely,

you'll need to build around the support posts and perhaps add support posts or bearing walls into your new stall plan. You can work this reshuffling of support into the picture if you move gradually, establishing the new support before you remove the old.

Whether you plan to handle the project yourself or use a builder, an exact cost figure will be very difficult to get a hold of. Whether it's in barns or homes, remodeling projects seem to invariably run over budget and far over time, no matter how well meaning the participants. Plan to build in stages to get a better grip on the financial side of things and see how you're progressing. Builders who work on a time-and-materials basis can be most helpful here.

Don't overlook the potential that existing structures may have in making your final plan a reality. There just may be a diamond in the rough behind those weeds or 50 years of cow manure.

A FIRE-DAMAGED VIRGINIA BARN

Albemarle Co. Barn Plan

This plan shows the reworked facility that was created from a burned-out shell in Albemarle Co., Virginia. A fire had destroyed the entire main section of a large old barn, leaving the adjoining apartment and an "L" off the main barn basically undamaged. The owner took advantage of the chance to restore and redesign the stabling, adding another "L" of stalls to the first. Working their way through the alphabet, they now have a handsome and extremely workable M-shaped stable that houses 15 horses.

While fire destroyed the original main structure, rebuilding it and remodeling the attached section maintained the unique style and usability prized by both the owner and those working in the barn. Old is not necessarily out-of-date or unusable in a barn. It may simply be tired, worn and in need of a facelift. (Albemarle Barn, Albemarle County, Virginia) Photo by M. F. Harcourt.

In this "before" shot, the bulldozer is just moving in to begin removal of the heavily damaged main barn section. Elderly outbuildings to the right were remodeled in the repair phase, pulling out tons of rotted wood and reorganizing the stabling plan. (Albemarle County, Virginia) Photo by Fritz Brittain.

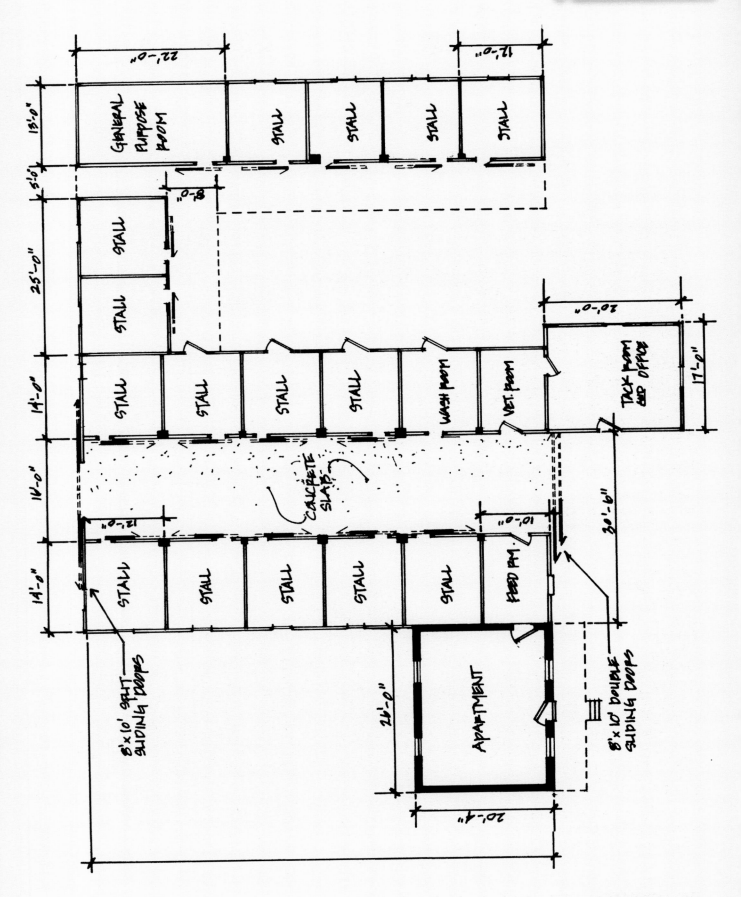

General Purpose Room

22'-0"
15'-0"

Stall
Stall
Stall
Stall

12'-0"

5'-6"

Stall
8'-0"
Stall

25'-0"

Stall
Stall
Stall
Stall
Wash Room
Vet. Room

20'-0"

Tack Room and Office

17'-0"

14'-0"

Concrete Slab

11'-0"

Stall
Stall
Stall
Stall
Stall
Feed Rm.

12'-0"

30'-6"

10'-0"

14'-0"

8'x10' Split Sliding Doors

Apartment

26'-0"

20'-4"

8'x10' Double Sliding Doors

The sight of the damaged barn is demoralizing, but offered a chance to replace the unusable section with a new piece that would complement the older buildings on the farm and throughout this historic area of Virginia. (Albemarle Co., Virginia) Photo by Fritz Brittain.

Here damaged siding has been removed and the lower sill, rotted out, has just been replaced by a new one after jacking up the structure slightly. At this stage concrete footers could have been placed, or, as was done here, treated lumber could be set in place. Next, the area will be backfilled to provide good drainage away from the barn. (Albemarle County, Virginia) Photo by Fritz Brittain.

To set up the overhang supports and provide a solid walkway in front of the old stalls, concrete footers and a drainage system down the center of the courtyard have been added. This construction is on the L attachment to the burned main barn, so a full drainage system was developed to avoid water seeping beneath

Most every barn needs refurbishing every 10 years or so. Checking for problems such as rotten boards and faulty plumbing, and adding fresh paint are mandatory for any facility. Photo by M. F. Harcourt.

Once the walkway edges had dried, herringbone bricks were added in for attractive and secure footing, and the support posts for the overhangs were bolted to the metal anchors preset in the concrete. These posts are now resistant to wind lift, and being slightly above grade they are less prone to moisture damage. (Albemarle County) Photo by M. F. Harcourt.

A DAIRY BARN REBORN

At February Farm in Lovettsville, Virginia, owners Trip Hoffman and Alan Van Wieren faced a major project when they looked over this 100-year-old dairy barn. A century's worth of cattle had left their mark on the place, and only a strong vision of the barn's possibilities transformed it into a scenic and functional horse farm. Pulling out the milking stanchions and the tons of concrete inside provided enough room for stalls and decent headroom despite the usual shortage of same in cattle barns. The milk storage wing, far left, was transformed into office space. Photo by M. F. Harcourt.

The sturdy old stone-and-frame barn structure provides historic shelter for horses, as well as phenomenal hay and general storage possibilities. The white door in the center of the loft wall slides open to allow a hay truck to back up for direct unloading. (February Farm, Lovettsville, Virginia) Photo by M. F. Harcourt.

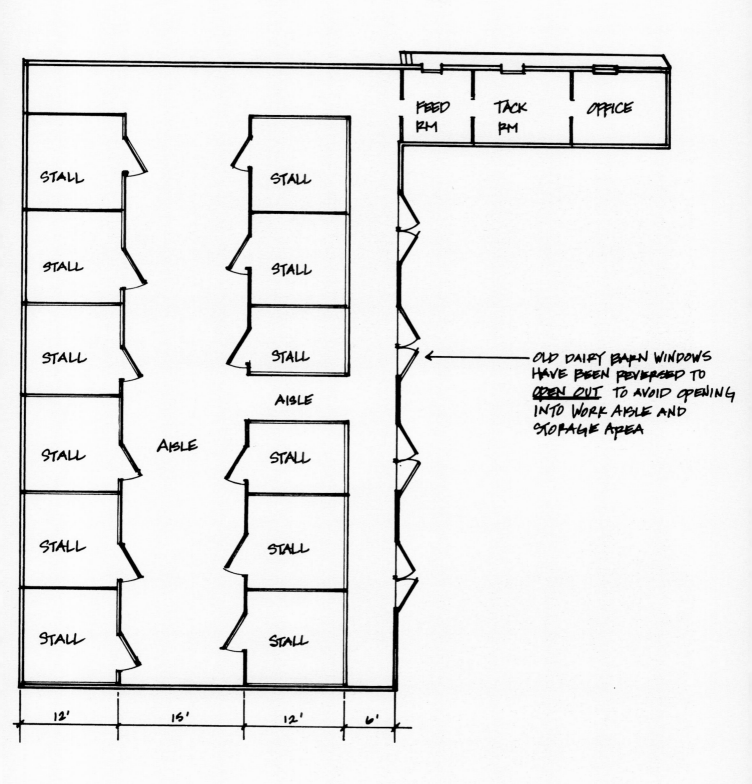

FEED RM

TACK RM

OFFICE

STALL

STALL

STALL

STALL

STALL

STALL

STALL

STALL

STALL

STALL

STALL

STALL

AISLE

AISLE

OLD DAIRY BARN WINDOWS HAVE BEEN REVERSED TO OPEN OUT TO AVOID OPENING INTO WORK AISLE AND STORAGE AREA

12'   15'   12'   6'

BELMONT BARN FALLEN ON HARD TIMES

This battered old Saratoga-style barn at Brandon in Spring Grove, Virginia, built in the 1920s, had been used for cattle for some years and had fallen into considerable disrepair. Owners Bob and Linda Daniel of Surry County, Virginia, took on the task of having it rescued and restored to daily use on their historic Brandon Plantation.

With many ideas of her own, Linda called on Equine Consultants, Inc., to help her think through her remodeling plans and offer suggestions as to additional facilities uses.

The original eight stalls were repaired to give three working stalls, a grooming stall with rubber mats adjacent to a tack room with a concrete floor and three stalls for storage and expansion.

By removing the outer walls on the back aisle, Linda created a full-length run-in shed serving her pasture. This allows her great flexibility in managing her horses. She can either stall them or simply let them stay out full-time knowing they can seek protection from the weather as needed. Photos by M. F. Harcourt

Support posts along the entire overhang had been rotted out, broken and damaged by loose cattle, so the decision was made to remove and replace all of them, including the broken concrete bases. (Brandon, Spring Grove, Virginia) Photo by M. F. Harcourt.

With the old posts pulled and the new ones firmly in place, the Daniels were ready to have the stalls worked over. Note the unusual ventilation system in this barn, small tilting windows at ear level on back walls and above each door. No stall had a view except when the Dutch door was open. This is one of the features that made this barn perhaps less ideal by modern standards, but interesting in terms of the prevalent styles when the barn was built. (Brandon, Spring Grove, Virginia) Photo by M. F. Harcourt.

Wood above about 3' was deemed worth saving, but sections were removed to allow installation of electrical outlets. Most of the doors were beyond repair, especially the bottom doors, due to rot and wear. (Brandon, Spring Grove, Virginia) Photo by M. F. Harcourt.

Where both doors had to be replaced, the original hardware was reused unless it had hopelessly degenerated. Here the original hinges are visible on the top door, but new steel ones are holding the lower door. Note the door frame itself has been salvaged, as has the transom above the door. Reconstruction of the doors included care to reproduce the narrow tongue-in-groove style of the old doors. (Brandon, Spring Grove, Virginia) Photo by M. F. Harcourt.

In order to refurbish the stall fronts, the doors, lower courses of wood and the sills were removed, and the soil beneath dug out. (Brandon, Spring Grove, Virginia) Photo by M. F. Harcourt.

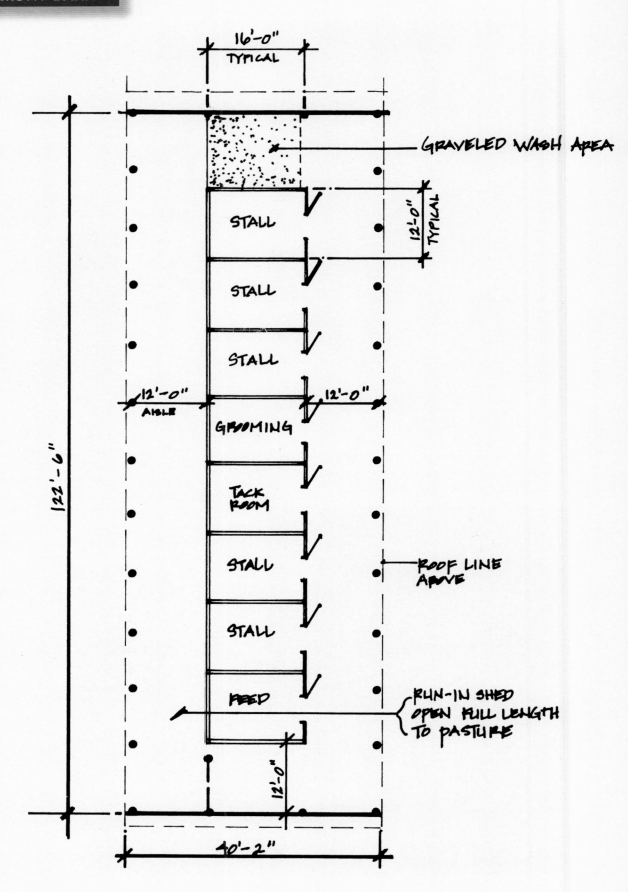

16'-0" TYPICAL

GRAVELED WASH AREA

STALL

12'-0" TYPICAL

STALL

STALL

12'-0" AISLE

GROOMING

12'-0"

TACK ROOM

STALL

ROOF LINE ABOVE

STALL

FEED

RUN-IN SHED OPEN FULL LENGTH TO PASTURE

122'-6"

12'-0"

40'-2"

DEERPATH'S 2 + 2 PLAN

This pair of plans shows a simple expansion of Deerpath Farm's barn from an attractive two-stall with a shed and combined tack and feed, to a four-stall package with a shed, center aisleway and a wash stall. Extending the roofline into a "T" shape allowed the change of floor plan, while adding greatly to the architectural interest of the small structure.

No "before" picture exists to show the barn as it originally stood, but in its remodeled state this is clearly a more substantial structure than a simple two-stall. Visible to the far right is an open storage or run-in area with two sides uncovered. The window beside it gives light and air to the tack and feed room. The standard, double-hung windows with inner grillwork are placed on two sides of the new stalls. (Deerpath Farm, Charlottesville, Virginia) Photo by M. F. Harcourt.

Cut into a forested hillside, this is a cozy facility with potential for further expansion. A fire extinguisher is just visible inside the entranceway to the right, and an amply louvered cupola adds to the upper ventilation provided by the louvered loft area at far left. (Deerpath Farm, Charlottesville, Virginia) Photo by M. F. Harcourt.

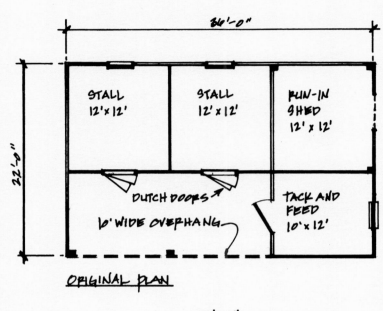

ORIGINAL PLAN

36'-0"

22'-0"

STALL 12'x12'

STALL 12'x12'

RUN-IN SHED 12'x12'

DUTCH DOORS

6' WIDE OVERHANG

TACK AND FEED 10'x12'

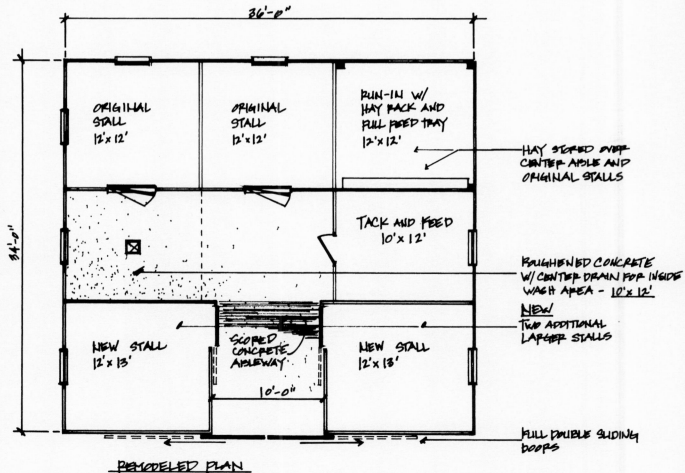

REMODELED PLAN

36'-0"

34'-0"

ORIGINAL STALL 12'x12'

ORIGINAL STALL 12'x12'

RUN-IN W/ HAY RACK AND FULL FEED TRAY 12'x12'

HAY STORED OVER CENTER AISLE AND ORIGINAL STALLS

TACK AND FEED 10'x12'

ROUGHENED CONCRETE W/ CENTER DRAIN FOR INSIDE WASH AREA - 10'x12'

NEW TWO ADDITIONAL LARGER STALLS

NEW STALL 12'x13'

SCORED CONCRETE AISLEWAY

10'-0"

NEW STALL 12'x13'

FULL DOUBLE SLIDING DOORS

This page is intentionally blank

This page is intentionally blank

Resource List

BUILDING RESOURCES

Agway Buildings
PO Box 4933
Syracuse, NY 13221
315-449-7061
www.agway.com

Barnmaster, Inc.
559 Floyd Smith Drive
El Cajon, CA 92020
619-441-9400 or 1-800-262-BARN
www.barnmaster.com

Barns by Gardner Incorporated
Steve Gardner
3833 West County Road 8
Berthoud, CO 80513
970-532-3595
www.barnsbygardner.com

Butler Manufacturing Systems
7400 E. 13th St.
Kansas City, MO 64126
816-968-3000
www.butlermfg.com

Cannon Ball, HNP
555 Lawton Ave.
Box 0835
Beloit, WI 53512
608-365-2121
800-766-2825, ext 121
www.cnbhnp.com
(gliding door tracks)

Chore-Time Equipment Co.
Triple Crown PVC, fencing
PO Box 518
State Road 15 North
Milford, IN 46542
574-658-4101
www.choretime.com
(automatic feed systems)

Country Manufacturing Co.
PO Box 104 C-5
Fredericktown, OH 43091
800-335-1880
www.countrymfg.com
(thermal buckets, manure spreaders)

Cover-All Building Systems
815 Wanuskewin Road
Saskatoon, Saskatchewan
Canada, S7P 1A4
800-268-3768
www.coverall.net

DT Industries, Inc
Exeter, Ontario
Canada NOM 1S3
519-235-1445

Equestrian Plastics
25215 Stanford Ave
Valencia, CA 91355
800-369-0303
www.magicplastics.com

Equustall (division of ACF Environmental)
2831 Carwell Road
Richmond, VA 23234
800-448-3636

Farnam Companies, Inc.
PO Box 34820
Phoenix, AZ 85067
800-234-2269
www.farnamhorse.com

Fortex/Fortiflex
800-468-4460
www.fortexfortiflex.com
(buckets, feeders, bay racks, etc.)

General Timber
625 Farmville Mine Road
Sanford, NC 27330
919-774-6213
www.generaltimber.net

Grain Systems, Inc.
PO Box 20
E. Illinois St.
Assumption, IL 62510
217-226-4421
www.grainsystems.com

Harrison-Banks
790 Boylston Street · Suite 23-F
Boston, MA 02199
617-236-1876
www.harrisonbanks.com
(architecture for the horse industry)

Humane Manufacturing Co
805 Moore St
Baraboo, WI 53913
608-356-8336
www.humanemfg.com

International Grating, Inc.
7625 Parkhurst
Houston, TX 77028
800-231-0115
www.seasafe.com

JHK Fencing
Jim Hamilton
875 N. Horace Walters Rd.
Raeford, NC 28376
910-690-1277

Lester Building Systems
Division Headquarters
111 2nd Avenue S
Lester Prairie, MN 55354
800-836-4439
www.lesterbuildings.com

Homes and Barns By Design
Linda Dreher
Architectural Designer
P.O. Box 865
Southern Pnes, NC 28388
910-245-7330

Linear Rubber Products, Inc.
5416 46th Ave.
Kenosha, WI 53144
800-558-4040
www.rubbermats.com

Louisana Pacific
414 Union Street, Ste. 2000
Nashville, TN 37219
877-744-5600
www.lpcorp.com

Midwest Plan Service
Agricultural Engineering
122 Davidson Hall
Iowa State University
Ames, IA 50011
800-562-3618
www.mwpshq.org

Mischka Farm
PO Box 2067
Cedar Rapids, IA 52406
319-362-3027
www.mischka.com

Morton Buildings
PO Box 399
Morton, IL 61550
800-447-7436
www.mortonbuildings.com

Nelson Manufacturing Co
3049 12th St., SW
Cedar Rapids, IA 52406
319-363-2607
www.nelsonmfg.com

North Carolina Building Enclosures, Llc.
Ellen Prince, Authorized Dealer
1822 South Glenbernie Rd
Suite 5
New Bern, NC 28562
800-232-8022
www.coverall.net

Northwest Rubber Mats
33850 Industrial Avenue
Abbotsford, B.C., Canada
V2S 7T9
604-859-2002
800-663-8724
www.northwestrubber.com

Peoples Pole Buildings
39 Erie St
Hubbard, OH 44425
330-534-1108

Petersen Company
1527 4th Ave, S
Dennison, LA 51442
712-263-2442
www.petersenmfg.com

PortaStall
PO Box 1426
Mesa, AZ 85211
800-717 7027
www.portastall.com

Ritchie Industries
120 South Main
Box 780
Conrad, LA 50621
641-366-2525
www.ritchiefount.com

Rockin J Horse Stalls
Box 896
Mannford, OK 74044
918-865-3366
800-765-7229
www.rockinjhorsestalls.com

Roka Agri Products
PO Box 2000
Peoria, IL 61656
800-447-2264

Rubbermaid Commercial Products LLC
3124 Valley Avenue
Winchester, VA 22601
540-667-8700
www.instawares.com/Rubbermaid-Guide.asp

Simpson Strong-Tie Connectors
Northern California Office
1450 Doolittle Dr
PO Box 1568
San Leandro, CA 94577
510-562-7775
800-999-5099
www.strongtie.com

Steelmaster Buildings
1023 Laskin Road, Ste 109
Virginia Beach, VA 23451
800-341-7007
www.steelmasterusa.com

Summit Flexible Products
12610 LaGrange Road
Louisville, KY 40245
800-782-5628
www.summitflex.com

VaFaC
217 Freedom Court
Fredericksburg, VA 22408
540-898-5425
www.horsestallsusa.com

Woodstar Products Inc
1824 Hobbs Drive
Delavan, WI 53115
262-728-8460
800-648-3415
www.wdstar.com

W-W Livestock Systems
3500 JFK Parkway, Suite 202
Fort Collins, CO 80525
800-999-1214
www.wwmanufacturing.com

INTERNET RESOURCES

If you have access to a computer, modem and a telephone line, you can find a number of excellent horse resources online, both from groups of horsemen who discuss equine subjects via the Internet, to the American Institute of Architects and a number of architectural design firm websites.

Commercial online services, such as America Online and Yahoo, as well as many privately owned websites, provide areas called forums or bulletin boards where you can leave messages on specific subjects (lots of "Barn-Building Questions" or "Stall Flooring Ideas" notes and responses appear over time). A great many horsepeople look over these posts and respond with answers, running from the sublime to the ridiculous. Chat rooms devoted to equine topics, which involve real-time discussions, are often supervised by one or more knowledgeable horsepeople who will usually welcome you as soon as they note your computer logging into their area.

In addition to the message and chat areas, archives where articles, stories, pictures and compiled messages on specific subjects (called "threads") are there for you to read or to download onto your own computer for later reference.

Online services offer access to the grand old man of the Internet, the Usenet News Groups, which include a group called "rec.equestrian" for horse folks of all flavors to exchange news, views and information. It's a lively, not-always-polite community, a virtual small town with strong personalities, hot topics and plenty of opinions running loose. When you first join, it's wise to sit quietly and familiarize yourself with the subjects, personalities and issues of the group before introducing yourself or asking what may be the 100th stall-floor question of the month. Ask something that's been beaten to death and you risk a harsh response, called a flame. Another point of "netiquette" is to always read the Frequently Asked Questions (FAQ) file before stepping into the fray.

Another type of networking is conducted via email, with mailing lists you can join such as Equine-L, a discussion group of several hundred horsepeople who serve as virtual pen pals. Rather than messages going one-to-one as with regular mail, each message you post to a mailing list is one-to-many, and many-to-one. Subjects range from training and feeding issues to detailed discussions of barn projects good and bad, with liberal references to additional sources of information and assistance.

Horse people have established some exciting and useful outposts on the Internet that incorporate text, graphics, sound and video. Using a type of software known as a browser, such as Netscape Navigator, Internet Explorer or an online services' proprietary package, a search using the words "horse" or "equine" at a site such as Google.com or Yahoo.com can produce incredible numbers of references to steer you to websites devoted to horse topics. The nature of the Web is such that most Websites will also include references to other worthwhile, horse-related Web pages—some that originate on the opposite side of the world. Each entry into the Web thus sends you down never-ending paths of information, like Alice in Wonderland going down the rabbit hole.

Some useful pages for barn and stable management information are noted below in standard Internet format, that is by their Universal Resource Locators (URLs):

Harrison Banks Equestrian Architecture:
www.harrisonbanks.com/

AEC InfoCenter: Architectural firms and services:
www.aecinfo.com

American Institute of Architects:
www.aia.org

NetVet's extensive listing of horsekeeping and health resources:
netvet.wustl.edu/horses.htm

Purdue University's Forage Home Page:
www.cas.psu.edu/docs/CASDEPT/agronomy/Forage

Oklahoma University's extension service:
osuextra.okstate.edu/dept/ansi/

Virginia Polytechnic Institute and State University's extension pages:
www.vt.edu/outreach

Planning and Architecture Internet Resource Center, University of Buffalo:
www.ap.buffalo.edu/sap/research/index.asp

Glossary of Building Terms

anchor bolt — a bolt for securing a machine, structure or part to masonry or other material, also called an anchor rod.

baffle board – a hinged board or panel placed in a wall or eave to prevent or allow air flow, depending on its position.

barn broom finish – roughened by the application of a coarse-bristled broom in any series of patterns, according to preference. Lengthwise strokes (and a sloped surface) encourage drainage.

batten board – a strip of wood used for nailing across two other pieces (as to hold them together or cover a crack) or a strip of wood used to strengthen or help seal a structure.

bearing block – a block of material acting as a bearing plate.

bluestone — often refers to $\frac{3}{8}$" to ¼ inch or less chipped stone of bluish color, used as arena or stall base, walkway surface, etc. Can also refer to any building stone of similar color.

braceboard – structural support, in this case bracing a padded bar against the weight of a breeding stallion.

breast bar – heavy bar, chest-height on a horse, that prevents forward movement. Often removable with steep pin fasteners.

bounce board – protective plywood sheeting along cinderblock will in breeding area, softening the wall so mare and stallion are not abraded should they kick or fall against the cinderblock surface.

bumper board – a padded 8"x8" board set horizontally along breeding chute wall offering a spacer to prevent the stallion's right foreleg from being pressed against the wall during service.

casing – wooden boards or strips used to encase structural and mechanical items in the barn, often for appearance, more often for safety.

cradle-like bracket – curved, iron bracket shaped to grip a telephone pole and welded in a Y configuration to a metal post or pipe.

cheater gate – section of fence designed to slide out of position easily and provide a gate-like opening.

collar beam – a tie beam connecting the rafters at a level considerably above the wall plate in a roof truss.

collar tie – a board used to prevent the roof framing from spreading or sagging.

concrete footing – concrete base for post or pole.

concrete necklace – concrete poured around the base of a pole or pole set in the ground.

concrete revetted wall – walls faced with concrete.

dowel pins – a headless smooth or barbed pin usually of circular section fitting into corresponding holes in abutting pieces to act as a temporary fastening or to keep them permanently in their proper relative position.

dropout rail – see *cheater gate*

fascia – a flat, horizontal part of a building having the form of a flat band or broad fillet.

fascia board – a horizontal board fascia covering the joint between the top of a wall and the projecting eaves.

fire stop – a member or material used to fill or close open parts of a structure for preventing the spread of fire and smoke.

French drain – a drain consisting of an underground passage made by filling a trench with loose stones and covering with earth, also called rubble drain.

gable-end fascia – a flat horizontal board used in a gable, the vertical triangular portion of the end of a building, from the level of the cornice or eaves to the ridge of the roof.

gable roof – a double sloping roof that forms a gable at each end

gambrel roof – a curb roof of the same section in all parts with a lower steeper slope and an upper, flatter one.

girder – a horizontal main member supporting vertical concentrated loads (as from beams).

girt – a heavy timber framed into the second-floor corner posts as a footing for the roof rafters in house building or a horizontal member running from column to column or from bent to bent of a building frame or trestle to stiffen the framework and to carry siding material.

grooved droppers – short posts of wood, fiberglass or plastic that support sections of high-tensile wire fence without being set in the ground.

gusset – a connecting or reinforcing plate that joins the truss members in a truss joint or fits at a joint of a frame structure or set of braces.

headers – a beam fitted between trimmers and across the ends of tail beams in a building frame.

hopper type opening vent – hinged boards for air-flow control (see baffle board).

in-line stretchers – tightening mechanisms for high-tensile wire fencing, set one wire on a multi wire fence and sometimes combined with springs.

joist – any of the small rect-angular-sectioned timbers or rolled iron or steel beams ranged parallel from wall to wall in a structure or resting on beams or girders to support the planking, pavement, tiling, or flagging of a floor, or the laths or furring strips of a ceiling. Also a similar timber supporting the floor. Or a stud or scantling about 3 or 4 inches in section. There are binding, bridging, ceiling and trimming joists.

louvered vents – an opening in a wall or ceiling for ventilation or cooling provided with one or more slanted fins to exclude rain and sun.

metal connector plates – a metal plate that connects two pieces of structural lumber at a variety of angles.

metal nail plate – a shaped metal plate with holes for nailing adjacent lumber sections together.

pipe fascia – a flat horizontal board made of pipe.

purlin – a horizontal member of a roof bracing the rafters and providing framing for roof-sheathing materials, also may resist wind uplift loads and gravity loads.

rafter – a beam that slopes from the ridge of a roof down to the eaves.

rafter plate – also referred to as rafter girts if builder is familiar with pole barn terms. These support the upper ends of rafters.

ridge – the line of intersection at the top between the opposite slopes or sides of a roof.

ridge beam – a beam at the roof ridge.

ridge girts – horizontal members running from pole to pole up on which the tops of the rafters sit.

ridge plate – a plate at the roof ridge.

ridge vent – a vent at the roof ridge.

sash – the framework in which panes of glass or other usually transparent or translucent material are set for installation in a window or door. Also the movable part of a window.

siding girts – girts used to attach siding to.

sill – a horizontal piece (as a timer) that forms the lower member or one of the lowest members of a framework or supporting structure. Also the horizontal member or structure at the base of a window opening serving to cover the wall at the base of the opening.

sill girts – girts used in a sill.

skirt trim – a part or attachment serving as a rim, border or edging.

sky belt – similar to a skylight, this band of transparent or semitransparent material runs along the top of a wall to provide increased natural lighting in a building.

soffit boards – boards used for the underside of a part or member of a building (as of an overhand, ceiling staircase).

splash board – a board or panel to protect against splashes.

starter board – vertical trim pieces at exterior corners of building.

stile – one of the vertical members in a frame or panel (as a door or sash) into which secondary members are fitted.

support girt – a girt used for support.

support poles – a pole used for support.

tongue and groove – a joint made by the tongue on one edge of the board fitting into a corresponding groove on the edge of another joint.

truss – an assemblage of beams, bars or rods typically arranged in a triangle or combination of triangles to form a rigid framework (as for supporting a load over a wide area).

twitch sticks – small (1- or 2') sticks used in corner or bracing sections of a fence, used as handles to tighten the wires that run diagonally between posts and prevent posts from leaning or shifting.

vertical studs – vertical support members for collar-tie support

wafer board soffits – inexpensive composite lumber product used under projecting rafter surfaces to close off eave area between the top of a wall and the roofing material.

zee flashing – metal or plastic material, Z-shaped in cross section and run horizontally, that deters water and drafts from entering where two plywood or similar flat surfaces are butted against one another, usually on a wall.

Index